Cleomar Alexandre Hirt

Innovative practices of the "good math teacher"

Cleomar Alexandre Hirt

Innovative practices of the "good math teacher"

In the Brazilian context

ScienciaScripts

Cover image: www.ingimage.com

This book is a translation from the original published under ISBN 978-3-330-75659-5.

Publisher:
Sciencia Scripts
is a trademark of
Dodo Books Indian Ocean Ltd. and OmniScriptum S.R.L publishing group

120 High Road, East Finchley, London, N2 9ED, United Kingdom
Str. Armeneasca 28/1, office 1, Chisinau MD-2012, Republic of Moldova, Europe
Printed at: see last page
ISBN: 978-620-8-27916-5

SUMMARY

DEDICATORY

I dedicate this monograph to my family, friends, work colleagues, our college bus drivers and everyone else who has been part of this journey in one way or another. I also dedicate it to those people who didn't believe in me, because in the most difficult times when you think about giving up, it's the arrogant words that kept me going. I dedicate all this success to those who helped me through the personal problems I faced, and to God, for being here at this moment. Thank you to everyone who has been part of this story.

ACKNOWLEDGMENTS

First of all, I would like to thank God for his strength and courage throughout this long journey.

I would also like to thank all the teachers who accompanied me during my undergraduate studies, in particular Prof.ª . Ma. Clàudia Maria Grando, who is largely responsible for the completion of this work, and for the fruits that we are reaping.

To my friends Eduardo, Rafael, Luis, Régis, Débora, Andréia, Bruna, Leomar, Rogério, Fagner, Ana, Carlice, Odirlei, Wagner, for all their support and complicity. Because even when they were far away, they were present in my life.

Thank you to everyone who, even though they are not mentioned here, contributed so much to the completion of this stage and to making me the Cleomar I am today.

I did it yesterday,

What's difficult I'll do today, what's impossible I'll do tomorrow

I'd rather die on my feet than live life on my knees. (Ernesto Che Guevara)

Dream and you will be free in spirit... Fight and you'll be free in life. (Ernesto Che Guevara)

Those who despise small events will never make great discoveries. Small moments change great paths. (Augusto Cury)

SUMMARY

To be excellent educators, it's not enough just to "be a teacher", we must have something more, something that transforms us and changes the lives of students. We worked together with innovative practices and differentiated methodologies, which were part of the theme of our research. It deals with the innovative practices of the good math teacher, from which we searched through the words of authors who have worked on the subject of the good teacher, the meaning, the basis of the good teacher, and thus we founded our theoretical practical research. Our objectives were to get to know their teaching practices, methods that increase the quality of teaching and what pedagogical innovations we are finding in math teachers. We tried to find out from the students' comments what the attributes of a good teacher are. By analyzing these three problems, we found the characteristics of a good teacher. Our research was initially based on a theoretical foundation, elaborated in order to understand what the main characteristics of a good math teacher are, we tried to formulate some questions to seek answers within our journey, to complete our research, we carried out a data survey, so that from what was indicated in the questionnaires applied, to the students of the last year of elementary and middle school, in the city of Pinhalzinho - SC. The schools involved were: E.E.B. Jose Marcolino Eckert, E.E.B. Vendelino Junges, E.M.E.B. Jose Theobaldo Utinzg, E.M.E.F. Maria Terezinha, Centro de Educaçao Objetivo. We have a total of eleven mathematics teachers. The questionnaire sought to make clear the student's idea of what the characteristics of a good teacher are. After collecting this data, in order to include everyone involved, we interviewed six mathematics teachers who were willing to answer our questionnaire, which sought to make clear what actions, what materials these teachers use. The aim was to make clear the position of the schools, students and teachers, so that we would have enough data to analyze the approximate profile of a good teacher, and discuss their attitudes in the classroom and how the lessons of this good teacher are developed. The data collected has been presented in tables and graphs, where we can see if there is a difference between the good elementary school teacher and the good secondary school teacher. We came to the conclusion that the good teacher is more open to conversation, that he is friendlier, that he also understands the situation of the students, that he works in a different way, presenting mathematics with more up-to-date concepts, that he enhances his lessons with some materials that aim to objectify the lessons in a more concrete way, that he charges the students on a daily basis, that he motivates and encourages teachers to search for knowledge even outside of school. More broadly, we can conclude that the research achieved 85% of its objectives, despite the difficulties in collecting the data. The results will be presented below, in a broad and organized way, subdivided into parts to make it easier for readers to understand, so that we can incorporate these qualities and use them to be considered good teachers.

Key words: Good math teacher, pedagogical innovation, differentiated teaching methodology.

INTRODUCTION

-A good teacher, a good start" is the title of the *jingle* in the advertising campaign of the Todos pela Educaçao Movement (financed by the private sector), launched in 2011 and broadcast nationally on television and the internet. The mobilization of the movement is to value the "good teacher". In this sense, it's not enough to be a teacher, you have to be a "good teacher". So we asked ourselves: what characteristics does (or should) a "good teacher" of mathematics have?

Along the same lines, in 2012 the RBS Group[1] launched a campaign entitled "Education needs answers". The campaign aims to raise awareness and mobilize the communities of Rio Grande do Sul and Santa Catarina to seek solutions to improve the quality of basic education in our country. Responsibility must be shared, so these initiatives also show the importance of rethinking our role as educators.

In mathematics classes, at different levels of education, we still see the teacher's performance anchored in strongly traditional pedagogical practices, centered on the transmission of information, whose list of didactic materials used is restricted to the textbook and the blackboard, practices that, for a long time, were recognized by society as good.

We *already* know that this model of teacher performance does not meet today's demands and has not been effective in ensuring that mathematics learning takes place in a meaningful way, but the repetition of practices from the models experienced during schooling and teacher training still prevails. As a rule, in undergraduate courses in mathematics, the concern with mediation for learning, with the use of various teaching materials, is discarded from the pedagogical work of teachers of the specific subjects of the course, and is left to teachers from the pedagogical areas: didactics, psychology, philosophy. What is achieved "is just a layer of pedagogical knowledge placed over the content learned, as if in learning content and form could be separated" (WACHOWICZ, 2009, p. 10).

1 Rede Brasil Sul de Comunicaçao, an affiliate of Rede Globo.

There are many difficulties encountered by teachers and students in the teaching and learning of mathematics.

Learning with understanding is a personal action, but by no means a solitary one, because it takes place in the social relationships established in different contexts. This process requires the learner to think for themselves, building themselves up as a social subject and not being subjected to a teaching process in which they remain passive. Thus, in the school context, the importance of the teacher is great. Their role is to make the path between mathematics and the student as meaningful as possible, in other words, to be a mediator of learning. To do this, it's not enough to be a teacher; you need to be a "good teacher" who challenges themselves to innovate and use a variety of teaching materials.

Faced with these concerns, the study we carried out was based on the following research problem: *What innovative practices of a good mathematics teacher are recognized by students in the final years of primary and secondary school?*

In order to delimit the study, we chose the following questions:

a) What are the characteristics of a good math teacher?

b) What pedagogical actions stand out in the work of a good math teacher?

c) What practices of math teachers do elementary school students find innovative?

d) What practices carried out by math teachers do secondary school students consider innovative?

e) Is the use of a variety of teaching materials in math classes considered by students to be a characteristic of a good math teacher?

Awareness of the inefficiency of mathematics teaching and, consequently, learning is the first step towards tackling the problem, but it is not enough. We need planned and theoretically grounded actions aimed at teaching that enables the establishment of contextualized relationships, the development of curiosity, autonomy, the pleasure of learning more and more, the spirit of investigation, considering mathematical knowledge not as an end in itself, but as a means of understanding and transforming the world.

In general terms, the research was carried out with the aim of analyzing the pedagogical practice of mathematics teachers, seeking to highlight the innovative characteristics of a good teacher, as indicated by students in the final years of primary and secondary school and by the teachers themselves.

[a]Specifically, we sought to characterize what this education professional should be like, surveying the aspects involved in the work of a good maths teacher based on what is indicated by students in the final years of elementary school (grade 8) and secondary school (grade 3), as well as by the maths teachers themselves, checking whether their actions are permeated by innovative practices, the use of a variety of teaching materials, or the occurrence of other variables delimited by the data collection, and whether or not there is a congruence between the indications presented by elementary school students and those of secondary school students, examining whether the use of a variety of teaching materials in mathematics classes is a determining factor for teachers to be considered good teachers.

In this way, research into the *innovative practices of the good math teacher,* especially considering the thoughts expressed in the discourse of primary and secondary school students and teachers, is justified by the importance of delving into what is being done in math classes, seeking a balance between what -should" be done - as proposed in the literature -, what "can" be done - given the conditions found in schools - and what actually "is" done - according to the involvement of each teacher - in order to get to know the reality of teaching action, reflect on it and collaborate to improve it, seeking to break unconscious practices of repeating a traditional teacher model, enabling a critical choice of action that involves the good math teacher.

The research report is organized into five chapters, the first three of which present the systematization of the theoretical study carried out. Chapter 1, entitled *Characteristics of the "good teacher"*, looks at the characteristics of what a good math teacher should be like, what are the practices and characteristics that should permeate the actions of these good teachers. In *The role of methods in the quality of teaching*, which is the title of chapter 2, we present, based on the ideas of the authors researched, the main methods that are part of the lessons of a good math teacher and their influence on the quality of learning. In the third chapter - *Pedagogical* innovations - we seek to clarify the meaning of the expression "innovation" that enters the school and its implications for classroom activities.

In the fourth chapter we describe the *methods and procedures* used in the research, explaining the subjects involved and the instruments used to collect the data.

In the fifth chapter we present the data collected, carefully analyzing it and organizing it into categories defined on the basis of the variables identified as relevant to determining the factors that produce interference in the proposed study. Finally, we present the *Conclusions*, which make clear the fruits of the research work, with considerations on the limits and possibilities identified.

CHAPTER 1: CHARACTERISTICS OF THE "GOOD TEACHER

School failure still appears today as one of the biggest problems in Brazilian education; it involves the failure of the individual (a student who doesn't learn and a teacher who doesn't teach), the fragility of the school institution and a socio-political system that is unable to implement actions to improve the educational situation.

Despite the widespread publicity about the improvement in the quality of education in our country, an example of which is the video produced by the Ministry of Education and Culture (MEC), entitled Brazilian advances in basic education[2] (2012), which indicates that this year Brazil reached (and in some cases surpassed) all the targets set in the IDEB (Basic Education Development Index), the situation is still not satisfactory. The most recent results show that the indices have increased in all stages of basic education (initial and final years of elementary school and also for secondary school). Despite this, the figures are still much lower than those of developed countries, which average 6.0. The causes of school failure are diverse, depending on the family, cultural, social and political context, and stem from education systems that fail to meet the diverse needs present in schools. The performance of the teacher, their preparation and involvement, is also a decisive factor in school failure or success. In the video cited above, the victory in improving the scores is attributed to the teachers.

As mathematics teachers, in this research we seek to carry out an expanded reflection that seeks to relate different factors, but focuses on the field of knowledge of mathematics education by analyzing the pedagogical practice of the mathematics teacher, seeking to highlight innovative characteristics for the "good teacher" of mathematics.

Next, we present our initial theoretical dialogue with different authors/researchers who have published on the compositional characteristics of the good teacher, looking in particular at the configuration of the mathematics teacher.

We will analyze the characteristics that teachers must have in order to be able, through

2 Available at: http://www.youtube.com/watch?v=pTx-vmHjsHs&feature=youtu.be.

their professional practice, to contribute effectively to student learning, to the construction of knowledge.

This is an old discussion. The Hungarian George Pòlya (1887-1985), a renowned mathematician and teacher (with an emphasis on Mathematical Analysis), dedicated the last 40 years of his life to the teaching of mathematics. In 1959, Pòlya published an article in the *Journal of Education,* University of British Columbia, Vancouver and Victoria, entitled "Ten commandments for teachers". This article was reprinted in Revista do Professor de Matemàtica (RPM, n. 10, 1987).

Pòlya indicates that he tried to systematize his opinions on the day-to-day life of a teacher in a condensed form, based on requests from secondary school teachers with whom he worked in training courses, stating them as *"ten rules, or commandments"*, which we present below:

Ten commandments for teachers

1. Be interested in your subject.
2. Get to know your subject.
3. Try to read your students' faces; try to see their expectations and their difficulties; put yourself in their shoes.
4. Understand that the best way to learn something is to discover it yourself.
5. Give your students not just information, but *know-how,* mental attitudes, the habit of methodical work.
6. Make them learn to make guesses.
7. Make them learn to demonstrate.
8. Look for aspects in the problem you are tackling that could be useful in the problems to come - try to discover the general model behind the current situation.
9. Don't reveal the secret all at once - let the students guess first - let them find out for themselves, as far as possible.
10. Suggest; don't make them swallow it by force. (PÓLYA, 1959 apud PÓLYA, 1987, p. 3-4).

In that article, Pólya comments on each of them, relating them in particular to the task of the math teacher, although he makes it clear that they can be considered for teaching in general.

In formulating the *commandments,* or *rules*, above, I had in mind the participants in my classes, secondary math teachers. However, these rules apply to any teaching situation, to any subject taught at any level. However, the math teacher has more and better opportunities to apply some of them than the teacher of other subjects. (PÓLYA, 1959 apud PÓLYA, 1987, p. 4).

We will now move on to an analysis of the Ten Commandments, looking at some of the characteristics found in these commandments.

In the first and second commandments, "*take an interest in your subject*" and *"know your subject",* it becomes clear that we need to know the content we are going to teach in depth. We can't teach something that we haven't mastered and that doesn't bring us interest and satisfaction, because what we don't have an interest in we won't be able to teach and awaken the students' "desire to know". When we get to this point, we have to be very careful. Even if the teacher doesn't like the content or doesn't know much about it, he or she must try to understand it better so that he or she can plant the seed of this knowledge in the minds of his or her students. As Pólya (1959) tells us, your interest will be proportional to your willingness to teach, in other words, the more you are interested in a certain subject, the more productively it will be reproduced in the form of knowledge. These conditions are necessary, but not sufficient, because "many of us have known teachers who knew their subjects but were unable to establish *contact* with their students" (PÓLYA, 1959 apud PÓLYA, 1987, p. 5).

-Try to *see your students' faces; try to see their expectations and their difficulties; put yourself in their shoes."* These are simple attitudes on the part of the teacher that make the moment of constructing knowledge a moment of pleasure. This commandment refers a lot to the teacher, because when the teacher expresses his or her arguments, they must be in line with the students' level of ability to understand. If the lesson makes the student feel sleepy, or feel unable to learn, there is something wrong with the planning or execution of the lesson. The teacher must idealize the lesson before applying it, foreseeing distortions in time, or the difficulty of understanding the way in which they intend to construct knowledge, put themselves in the students' shoes, feel the sensation that the student is feeling, and make this sensation pleasurable. The teacher's degree of skill is related to his or her experience and willingness to innovate, proposing new practices and strengthening the interaction between teacher and students.

In the fourth commandment, *"understand that the best way to learn something is to discover it yourself",* Pólya (1959) meant that the teacher must instigate, provide tools, so that the student reaches the problem and sees what is proposed, -that *'active'* learning is preferable to *passive*, merely receptive learning. The more active, the better the learning." (PÓLYA, 1959 apud PÓLYA, 1987, p. 5).

"Give your students not only information, but know-how, mental attitudes, the habit of methodical work", in this fifth commandment, we are already talking about the way in which the teacher is teaching his students, often the way in which one teaches is more valuable than what is being taught, the teacher himself needs to create and improve teaching methods that are flexible to be used in various classes, to make students learn how to learn. It's not just a question of the teacher making the students accumulate information, but of making them develop the skills to use and deal with this information and to be able to solve problems by mobilizing what they have learned and, in this way, effectively constructing knowledge.

> *Know-how* is dexterity; it is the ability to handle information, to use it for a given purpose; *know-how* can be described as a collection of appropriate mental attitudes, *know-how* is ultimately the ability to work methodically.
>
> In mathematics, *know-how* is the ability to solve problems, construct demonstrations, and critically examine solutions and demonstrations. And in mathematics, *know-how* is much more important than the mere possession of information (PÓLYA, 1959 apud PÓLYA, 1987, p. 6).

The sixth and seventh commandments, *"make them learn to make guesses", "make them learn to demonstrate"*, are already intertwined with the previous one, which stated that the student will discover ways of finding solutions on their own, bring the content, show it, but let the student become familiar and work with the content, give the students the opportunity to argue, give them the opportunity to argue, to make guesses, to build up a more scientific mindset from suppositions or assumptions, to develop an intuitive mindset on their own, always based on what they have already learned throughout their lives, "first conjecture, then prove - this is how discovery proceeds in most cases"

(PÓLYA, 1959 apud PÓLYA, 1987, p. 6). 6).

Pólya (1959) highlights the importance of developing "plausible reasoning", which encompasses the ability to make reasonable, judicious guesses, as well as "demonstrative reasoning" as a way of logically validating a statement. These two skills can be taught and learned, respecting the levels of education, going far beyond the training of mere mechanical processes of arithmetic, algebraic or geometric calculation.

The eighth commandment, *"look for aspects in the problem you are tackling that may be useful in the problems to come - try to discover the general model behind the present concrete situation",* shows once again that "know-how is the most valuable part of mathematical knowledge, much more valuable than the mere possession of information" (PÓLYA, 1959 apud PÓLYA, 1987, p. 7). In order for the teacher to use the tools to build generalizations of procedures and reasoning with the students, it is important to show and link them to real situations; every student likes to know where such content is used. This teaching tactic could be a practice for all teachers to innovate, to familiarize themselves with this idea, and also to seek out a little more information to pass on to their students. The way in which a question is approached is a complex job which depends much more on the student than on the teacher, but the teacher can give indications of how to arrive at these techniques.

Hence the rule: *Look for aspects in the problem you are addressing that may be useful in the problems to come - try to discover the general model behind the present concrete situation.* (PÓLYA, 1959 apud PÓLYA, 1987, p. 7).

Finally, Pólya says *"don't unlock the secret at once - let the students guess first - let them discover it for themselves, as far as possible" and "suggest; don't make them swallow it by force".* These commandments express the practice of teaching, the way in which the teacher constructs knowledge together with the students. The teacher must make his or her lessons a space for discovery, for the construction of reasoning, which are interconnected throughout the school journey, the teacher must find a balance in the way knowledge is transmitted, The teacher must find a balance in the way he or she transmits knowledge, knowing facts such as letting the students walk a little on their own, because this instigates the ability to produce knowledge, and this is one of the main objectives of the school journey, to make the students the builders of knowledge themselves, with the mediation of the teacher.

Especially when explaining solutions, which can have several ways of being done, try to suggest and expose the variations, because each student is different, and each one finds one way or another easier or more difficult, and it is in these differences, between one and the other, that many students understand the essence of the construction of reasoning.

After analyzing the ten commandments for teachers one by one, we can build a more general analysis of the subject. Pólya, back in 1959, which is the date of the original publication of the text in which he presents his commandments, highlighted very well small actions that a good teacher should carry out, or ways in which they should act in class, so that learning becomes a much more exciting moment. A good teacher won't necessarily have all these skills, but if they are aware of some of them, as has already been said, experience will bring the others. The good teacher is the one who tries to be understood by the students and also tries to understand the students. To make it clearer, we can say that to be a good teacher we have a recipe with a few ingredients, but what makes it work is love, affection and interest in teaching and mediating teaching, dosing the content in the right way, always developing a way with each class that satisfies its profile because each class has its abilities and its flaws, it is the teacher's job to nurture the abilities and reduce the flaws.

After Pólya, who dared to talk about the role of the teacher at a time when didactics was not considered important, when teaching was based on authority and not on dialog, there have been many studies and indications regarding this complex issue.

We can see that there are many elements that contribute to or detract from the teacher's performance: the way in which knowledge is transmitted (which presupposes the teacher's mastery of the content, mastery of teaching methods and techniques); the characteristics he or she must possess in order to be an extremely important figure in the students' lives (believing that everyone has the capacity to learn, valuing what each student excels at, but also trying to encourage the development of other skills and abilities).

The teacher-student relationship is also one of the reasons for good learning, as Cunha (1991) describes in a report following his field research with 2nd and 3rd grade students (secondary and higher education), relating to the elements of a good teacher, emphasizing that:

> [...] it would be difficult for a student to name a teacher as good or the best in a course without them having the basic conditions of knowledge of their subject, or the ability to organize their lessons, as well as maintaining positive relationships. However, when students verbalize why they chose a teacher, they emphasize the affective aspects (CUNHA, 1991, p. 145).

Cunha (1991) points out that in many of the students' reports obtained from the research, we find expressions such as "he is a friend", "he is understanding", "he is like us", "he cares about us", "he is available even outside the classroom", "he is fair". These ideas show us that, when emotional bonds are established between teachers and students, there is an even greater awareness of the coherence of love for teaching, the pleasure of teaching for the benefit of the people around us on a daily basis.

For Cunha, there is a balance between the teacher's being and doing.

> A teacher's behavior is a whole and certainly depends on the worldview he or she possesses. I don't know to what extent it's important or possible to classify teachers. Not least because they too, as a result of social contradiction, don't always behave in a linear way that is totally consistent with a philosophical current. It is undeniable, however, that man's way of being and acting reveals a commitment. And it is this way of being that once again demonstrates the non-neutrality of the pedagogical act (CUNHA, 1991, p. 146).

In this research, we will try to find out what the characteristics of the so-called "good teachers" are and what they are like, so that we can question the possibility of developing them in ourselves, or in future teachers. This topic is wide-ranging and can affect a multitude of students at different levels of education. By building a small idea, however simple it may be, it will still bring some change to these thousands of people who depend on these teachers to build a professional future.

The teacher-student relationship is not restricted to the emotional bonds created, but also relates to the way the teacher deals with the content, the teaching skills they develop and the value given to the pleasure of learning. We can see this because the students, in their answers to the questionnaires carried out in Cunha's research, talk about their satisfaction with teaching and learning, saying:

- *- I choose this professor as the best because of the way he makes us think, not by presenting the theoretical content as a finished truth, but by questioning it";*
- *"What I like about Professor X is that he's always ready to answer our questions, he even encourages us to have questions... ";*
- *"Teacher Y is the best because he conveys his love of mathematics to us. He shows us the pleasure of learning..."* (CUNHA, 1991, p. 146-147).

Another aspect that stands out in Cunha's research is methodology. Classroom practices show when teachers believe in their students, in their potential, when they try to involve them and make learning enjoyable, promoting a high level of student satisfaction. The students surveyed by Cunha cite that -among the characteristics of their best teachers are 'making classes more pleasant and attractive', 'stimulating student participation', 'knowing how to express themselves in a way that everyone understands', 'inducing criticism, curiosity and research', 'making students participate in teaching' etc...". (CUNHA, 1991, p. 147).

The teacher must polish the student as if they were a rough diamond, but always seeking to give them something more, a special touch. This is how the good teacher builds their lesson, their knowledge, and the student *can already* see whether the teacher is willing to give a good lesson or not. The good teacher will use all possible techniques before giving up on the student who is no longer looking for anything, or can no longer find anything to hold his or her attention in class.

In the school environment, you don't find many of these figures who, most of the time, are idolized for their ability to teach and maintain a friendly attitude towards their students, as Oliveira (2007) tells us.

Every student creates an expectation of their teachers, supported by social and cultural elements, but, as Cunha (1991) points out, it is not generally linked to the teacher's political position. The teacher's ideology, which is evident in the way he or she acts in the classroom and which is also increased outside the school institution, in the wider social sphere, is taken into account to the extent that this ideology reflects on the construction of knowledge, on the enjoyment of teaching. Still,

> It's important to say that students don't point to the so-called "good" teachers as the best. On the contrary. Students value teachers who are demanding, who demand participation and tasks. They realize that this is also a form of interest, which is linked to everyday classroom practice (CUNHA, 1991, p.148).

The act of the teacher teaching can be compared to a performance by a magician: if it is performed in an entertaining and correct way, the audience will certainly enjoy it and will always remember it, but if the magic is flawed, surely, after a short time, no one will remember how it was performed and, in the case of learning, it follows the same line: if the teacher is not good or, at the very least, if the teacher has not made an effort in the elaboration of his teaching method, in his planning, learning will not take place.

Oliveira (2007) and Aurich (2011) agree that the teacher must be the student's friend, in such a way that the student can feel free to put forward their ideas, to argue, to question and to answer their questions; to have the freedom to talk to the teacher, because they will be the teacher for the rest of their lives. The good teacher is the one who carries with them the desire and satisfaction of seeing the student learn, and does their utmost to ensure that this happens, because the good teacher seeks, within the problems they face, to build a road to knowledge, because if the teacher lives the problem they will know which action will bear good fruit.

In a survey carried out by Oliveira (2007), using a questionnaire with 2 open and 31 closed items, she obtained a large amount of information in an attempt to understand the teacher's performance in the classroom, teaching methodologies, the teacher-student relationship and assessment practices. In this survey, she highlights some of the points that students most often mention about a good teacher, one of which is the maintenance of friendship and collegiality between teacher and student, as well as the outgoing teacher, there is also the hard-working, dedicated teacher, basically students refer to a good teacher by their profile as a human being.

The good teacher is always ready to face any situation, because he or she is prepared, has planned what he or she is going to work on in various ways, and is, above all, enthusiastic about teaching. As Pólya says

> [...] teaching is a process that has lots of little tricks, highlighting that this process has a lot in common with theatrical art. For example, when a teacher has to present his class with a demonstration that he knows in detail because he has presented it several times in previous years in the same course. In reality, he may not even be that enthusiastic about the demonstration anymore, but that shouldn't be shown to the class [...] In this way, it's better to pretend to be enthusiastic when starting the demonstration, to pretend to have brilliant ideas when developing it. Pretend to be surprised and elated when the demonstration is over. The teacher should act a little for the sake of his students who, in some cases, may learn more from his attitudes than from the content presented. (1981, apud OLIVEIRA, 2007, p. 3).

The way the teacher presents him/herself in the classroom is also very important, because the teacher has to reassure the students, he/she has to show that he/she is confident in order to reassure the students who often consider the math teacher to be a "seven-headed monster". The tone of the teacher's voice must lead the students down the right path, showing them what is hidden among the rocks to be faced. The act of teaching comes directly from the teacher and much of what counts is in the attitude, in the way of showing things. The good teacher should seek to be a master of oratory, speak in different ways, develop strategies that help the student to reach knowledge, but in a diversified way; these attitudes are essential for anyone who is or who wants to become a good teacher.

We emphasize that we are not looking for a single, infallible model, but for elements that are socially considered important to define the competence of a "good teacher", as Cunha points out:

> I'd like to clarify, their competence is seen here as the idea of a socially located role. There is no intention of defining a finished, neutral model. It is very much linked to the idea of a social role. In this respect, it is worth remembering Heller (p. 87) for whom -... the idea of a social role is not born casually, or out of nowhere, but is the result of numerous factors in everyday life, given before the existence of this function and which will continue to exist when it has been exhausted" (CUNHA, 1989, p. 89).

The teacher is seen by society as a being who will inspire the students, in other words, the teacher already comes into the classroom with this pressure. According to Cunha (1989), teachers are already overloaded with activities and are still given the responsibility of building morals and ethics into the character of their students.

Nowadays, this is being demanded a lot by parents, but without them being concerned about everyone doing their bit. The teacher has become the school's do-it-all, the teacher must make ideas flow in a positive and correct way, the teacher is placed as the only person responsible for education, thus making their role ever greater and, often, the situation becomes untenable to the point where the teacher loses all prestige with their class.

The teacher represents much more than the mediator of knowledge, the teacher must construct new knowledge, starting from the daily reflection of their teaching practice and the situations they experience, and not just reproducing the same actions over and over again. Moysés points out that the teacher produces much more than is imagined: "the act of teaching is something that changes constantly, and can work for some and fail for others, because the universe of education is something very large and the teacher is merely an object in this context" (1994, p. 46).

We're trying to build up an idea of what an ideal teacher should be like so that we can pass it on to our colleagues so that, if we don't reach this ideal figure, we can at least get close to it. Analyze how this good teacher transmits the content, what the focus of his or her lesson is, how he or she approaches knowledge, because what is at stake is the teacher's teaching philosophy. To look closely at what makes this or that teacher different, what makes him or her good, what characteristics or differences he or she possesses, because in common all teachers have the duty to teach.

As a researcher who is studying for a degree in Mathematics, it is important to better understand the methods that good teachers use to transmit the content effectively and with more enthusiasm, in a way that is contagious, that mobilizes psychological and cognitive aspects of the students and makes it easier for them to learn. In this sense, we present another aspect of our reflection in this research related to teaching methods.

CHAPTER 2: THE ROLE OF METHODS IN TEACHING QUALITY

The art of teaching was born more than 400 years *ago*, in the hands of Comenius (1592-1670), through *Didàctica Magna - a treatise on the art of teaching everything to everyone*. At that time, ideas began to emerge, such as the varying learning abilities of each individual and the fact that knowledge should be taught in different ways. We can compare this to the present day. These transformations that took place at the end of the 16th century and the beginning of the 17th century are very much reflected in the construction of the ideas put forward today for an improvement in education, everything is being restructured to adapt better to the new times, to new ways of thinking, even in the way society is constituted.

Moysés (1994) tells us that the challenge of teaching today is greater, because the school is no longer at the center of students' attention. But for everything to fit together perfectly, the main thing is that the teacher must master the science he or she is working with, having in-depth knowledge of its concepts and, as an educator, knowledge of psychology, didactics, in short, the pedagogical knowledge needed to teach. The teacher must be flexible in the way he or she teaches, taking into account the student's social environment and the fact that not all students learn in the same way, because not everyone is the same.

Knowledge is not the result of just one action, but of a collective of organized actions.

Various elements are related to the learning process, such as the organizational and administrative structures of educational institutions, the composition of classes, the involvement of family members, the distribution of school days and times, among others. However, these elements, which are important, do not directly affect learning. [Learning happens when the three main components - teacher, student and curriculum - interact with each other, producing a spontaneous intellectual or artistic combustion (RENZULLI, 2001 apud OLIVEIRA, 2007, p. 1).

The essence of a good teacher lies in various aspects: the teaching tools he or she uses, the elements of didactics, the desire to be a committed and qualified professional. This idea makes the teacher an "educator", because educators have the ability to see education with different eyes, to perceive it in a committed way or with a friendlier feeling about how to be a good teacher, how to be a "superhero" within our education system so full of problems.

Being a good teacher requires much more than just doing the simple things. Being a good teacher, according to Perrenoud (2000), means varying teaching tactics, seeking satisfaction in the eyes of the students, but for this to happen, a lot of effort is needed, which teachers are often unable or unwilling to expend.

To start improving their work, teachers simply need to organize their lessons better and increase them with different activities, as Rizzotti (2005) says.

The good math teacher sticks to new tools, as Aurich (2011) tells us, they can be new pedagogical practices (new for the teacher). It sounds a bit simplistic, but if the teacher uses different tools for teaching, he or she can make lessons more dynamic and much more learning will take place. There are many suggestions for teaching practices and materials that teachers could use to enrich their lessons, but what is often lacking is a link between the pedagogical knowledge produced in research and action in the world of mathematics teaching.

It's not enough for a teacher to have in-depth knowledge and mastery of the content they have to teach their students, that's one aspect; the other is mastery of teaching techniques, combining content and form. Oliveira (2007) talks about the lack of combustion between these two aspects, i.e. from the point of view of a pedagogue, mathematics should be taught in a more elaborate way, interspersing games, play, problem-solving, modeling, different materials, experiments, simulations, software, with the knowledge to be developed.

Aurich (2011) has been increasing the construction of the figure of the good teacher and also links it to their academic training. As a good math teacher we would like to construct ourselves, in addition to proposing the meaningful lesson of the good teacher, in her search for answers she sets an important milestone when she says that she does not intend to find or prescribe the good teacher, but rather to approach him in a broader way in the eyes of all who want to follow suit. According to Aurich (2011), this involves much more than an insane search for the competent teacher or for an ideal teacher who fulfills a certain list of skills or competencies, according to her, there is no recipe for becoming a good teacher, but rather, there are some attitudes to be followed, and countless possibilities for being a teacher, and acting like one.

Pólya (1959) criticizes the way teacher training is carried out, pointing out that teachers' university studies help them very little to follow the ten commandments of the teacher that he proclaims, highlighting the gap between technical training and pedagogical training.

> Very often, future teachers leave secondary school without any knowledge or with a hesitant knowledge of secondary mathematics. Where and when should they learn secondary mathematics?
>
> He takes a course offered by the Mathematics Department on more advanced topics. He finds it very difficult to adapt to and pass the course because his knowledge of secondary school mathematics is inadequate. He can't relate the course to his secondary mathematics. On the other hand, he receives a course offered by the Department of Education on teaching methods. This is offered according to the principle that the Department of Education only teaches methods, not content. Our future teacher may get the mistaken impression that teaching methods are essentially related to inadequate knowledge, or ignorance, of the content. In any case, their knowledge of secondary mathematics remains marginalized.
>
> [...] The teacher is exhorted to do many beautiful things: he must give his students not only information but *know-how,* he must encourage their originality and creative work, he must make them experience the tension and triumph of discovery. But what about the teacher himself? Is there any opportunity in his curriculum for independent work in mathematics, for acquiring the *know-how* that he is expected to pass on to his students? (PÓLYA, 1959 apud PÓLYA, 1987, p. 9)

Faced with the deficiencies in initial teacher training that we know are still present in undergraduate courses, continuous improvement must exist, especially if the teacher is focused on trying to give the best to their students, but some elements are often not part of the teacher's profile, which requires commitment in the search for new educational practices. In order to be competent, "the teacher must master the content of their subject, be able to use various appropriate instructional techniques to transmit this related content and also have the ability to develop a romance with the subject they teach" (OLIVEIRA, 2007, p.12). This highlights another element: passion, for the content and for teaching. This feeling will make the challenges enjoyable.

A crucial question raised by Oliveira (2007) in relation to good mathematics teaching is: do we need a good teacher or good students? The practice of teaching is a two-way street; learning will only occur if the student is involved and willing to participate in the process. The student is the main element towards which the teacher's action is directed. Both teacher and student must be mobilized for the construction of knowledge. The teacher cannot know for the student and neither can the student be responsible for the process of pedagogical mediation, so each has their own well-defined role.

The role of the teacher implies a responsibility for constant improvement so that they are capable of effective, efficient and effective pedagogical action. This professional must also be competent to train citizens who are fit for social and professional life. It is therefore worth emphasizing the importance of good training for teachers, as this will be reflected in their daily teaching practice (PERRENOUD, 2000). Teachers can't just be the ones who deposit and transfer knowledge to their students, but must take on a new stance, that of knowledge mediator (OLIVEIRA, 2007, p. 2).

The good teacher is the one who recognizes that, nowadays, the learning process must be differentiated, because as Freire describes, "it is the subject who builds, but from the social relationship, mediated by reality" (apud OLIVEIRA, 2007, p. 2).

The teaching techniques applied to many students leave something to be desired, because it seems that thc teacher is depositing knowledge in the students' heads and, in most cases, thinks that knowledge is like a "seed" that will grow on its own. Oliveira (2007) tells us that the student must like mathematics and the good teacher must mediate this so that the student can see mathematics in a more attractive way.

The profile of the teacher of the new millennium is undergoing a change, both in their skills and knowledge and in their competencies, so that they don't stop fulfilling their role and never stop "learning to learn".

The idea that teaching is not a science, but an art, has already been expressed by several people. Several studies have been carried out to try to unravel the mysteries of this art (OLIVEIRA, 2007, p.3).

A good teacher should set up his or her lessons in different ways, varying the lessons, interspersing videos, excursions and different books, showing enthusiasm, excitement, liveliness and a sense of humor, showing students that they can go further than they think in the classroom, planting dreams, It's about planting dreams and watering them on a daily basis, and debating ideas that emerge on a daily basis, generating discussion of knowledge, in this way building much more lasting knowledge, because direct intervention in the construction of an idea can fix it in a much more lasting way.

Cunha (1996) talks about teacher failure in the face of inadequacies in the search for new practices: failing is human, but seeking to correct mistakes takes precedence over any other attitude. The dissemination of knowledge is a complex process of building skills, and these skills must be present on both sides, in the teacher and the learner.

Knowledge is not a product ready to be consumed, as Perrenoud (2000) points out.

Knowledge is raw material which the teacher must reshape in such a way that his or her students begin to unravel this knowledge, which is not ready, but can be discovered and reconstituted in a more personal way, perhaps in the form of research, which is a form of learning that is much in demand today, as it instigates the search for knowledge, for its own polishing, as well as exposure to error, which is also part of a subset of learning. Perrenoud (2000) points out that research is a very comprehensive object of study, as it can lead people to learn in a more differentiated way and not in the mechanical way that is common in the current education system.

As Perrenoud (2000) points out, the teacher's aim should be to complete, in the classroom, what the student cannot find elsewhere. He also defends the idea that school failure is a reality manufactured by our own society, that the naturalization of school failure contributes to building up a pessimistic mentality in students' minds about school.

As teachers we can make a difference, even in the face of such a complex and sometimes even discouraging picture of Brazilian education.

There are successful experiences whose difficulties have been overcome and can be considered full of innovation.

Next, we try to characterize pedagogical innovation, seeking to deepen it so that we can later identify and analyze the innovative practices of the "good teacher" of Mathematics, which is the focus of our research.

CHAPTER 3: PEDAGOGICAL INNOVATION

When we talk about innovation, we need to understand a little about it, how it happens, how it evolves, in what sense to innovate, to seek in current practice the essence of teaching as a knowledge to be reconstructed by the teacher, by each student, whether an academic or a primary or secondary school student.

For Demo, educational innovations are "initiatives that enhance the opportunity to learn well" (2012, p. 14).

> [...] I see educational innovation as the chance to offer those who need it most a genuine opportunity to learn well, both in and out of school, preferably with digital support.
>
> I don't consider the reforms of this outdated instructional system to be educational innovations, such as: literacy in three years, the current UCA, the Comprehensive School (a caricature of the Full-Time School), the pedagogies that want to be "face-to-face", "democratic management" (because there is nothing to manage as long as there is no minimally acceptable learning in schools), the increase in classes, etc. (DEMO, 2012, p. 15).

The focus should be on learning, making teacher and student "authors" and not mere reproducers:

> [...] I emphasize that I usually revolve around the challenge of "learning" for the student and the teacher, because I consider one of the most intrinsic evils of the current system to be "instructionalism": pedagogies of reproduction. Since the most promising opportunities for life and the market are centered on the ability to produce knowledge (Amsden, 2009), I assume that the student comes to school not to listen to a lesson (generally copied and made to be copied), but to become an author, individually and collectively (DEMO, 2012, p. 7).

Let's look again at the teacher and his or her practices: what has he or she failed to do in his or her lessons? And what should he try to propose, to engage in dialog, so that his classes become more attractive? These are difficult questions to answer, but many would do so easily, assuming that the teacher built more diversified lessons and included some resources, and that would be the end of the good teacher with his innovative practices. These practices are much more than simple planning actions, they are real paradigms to be broken in society itself. A teacher who includes in his teaching didactics and annual lesson plans elements that propose recreating knowledge, based on situations that are different from the usual ones, will often be considered a denatured teacher in the eyes of others. Because innovation also requires many sacrifices.

Demo criticizes the organization of the Brazilian education system, arguing that it was born to fail:

> This became even clearer to me with the National Education Plan (PNE), a piece of work surrounded by a certain amount of civic goodwill, but which is retrograde because it continues to bet on this "education system" - it seeks reforms, adjustments, remedies, such as increasing the number of school days, for example, while ignoring stubborn problems such as those of the mere teaching university, the clamorous and chronic shortage of science and math teachers, school instructionalism, literacy in three years (which enshrines automatic progression and further traps the poorest), prehistoric pedagogies and degrees, inclusive digital literacy, waving "democratic management" as a consolation for this "poor thing for the poor". Democratic management is too important a challenge for education - it's the means and the end - to be squandered in an unsuitable system: as the student doesn't learn, there's nothing to "manage", and even less does this school have any relation to "democracy". In my practice, the first step is to question this system, looking for alternatives that, in law and in fact, produce learning, in other words, a learning system, not a teaching system (DEMO, 2012, p. 7-8).

In our country we live on the "basis of the way", of "let it pass", "it will get better". The problem is widespread, from the faulty organization at the top of the country's educational planning, until the bomb hits the poor teachers who are easily blamed for many school failures. Often teachers follow orders and don't carry out their own wishes, becoming mere puppets in the classroom.

The teacher must construct knowledge in order to at least show his students the way, so that the students themselves construct their own learning, and not just be taught automatically, making them just another student on this journey.

> Learning requires research, self-crafting and production, individual and collective authorship, active participation, virtues that pedagogies and degree courses do not practice or disfigure, generally because of their non-author professors. I understand that without research there can be no class: it will be mere copying. That's why it doesn't make sense to have a university that merely teaches, because it's an unnecessary institution: "trema" (it doesn't train). (DEMO, 2012, p. 9).

We can put research as one of the first elements to be analyzed, as an innovative teaching practice, many teachers use it, the vast majority in the wrong way, forcing their students to look for information that doesn't interest them because they don't know much about the subject. That's the question: how can we include research in the school process without making it a stifling practice for students? In research, students look for information on their own, and we can mention how they will know if this or that piece of information is really related to the context. However, when we discuss with students which sources they should trust, there should be a search for ideas that diverge and converge in order to have several options for opinions and content.

The practice of research in primary schools can be centered on motivating and meaningful problematizations that the teacher is responsible for organizing and proposing, with the students, individually and collectively, being the authors of the elaborations made on the basis of the proposed problem.

> The challenges are many, starting with the teacher, who needs to let go of the classroom as the central pedagogy. The teacher's role is to organize the student's productive work, with attractive and realistic problematizations of the curricular content, so that instead of attending class, the student researches under the guidance of the teacher (DEMO, 2012, p. 13).

The knowledge acquired through research is of the utmost importance, because it is a construction, an intellectual elaboration by the student himself, something that is rare at the moment in education. The student researcher can be anyone who searches for information in different sources, diverging and comparing them with each other and constructing their own point of view, or their own knowledge. Research is learning how to learn. Researching for its own sake is an element that should, in Demo's (2012) view, be included as a learning practice, because in this way it is necessary for the student to seek knowledge on their own and with the help of the teacher.

One issue to be studied is that not everything can be put to students for research, the most complex content would certainly not be the most suitable to receive this type of treatment, research is an excellent systematization of information, as long as it is well organized, and built with the help of a teacher who guides and shows certain options and varied paths.

The evaluation process can also be considered a facet of pedagogical innovation.

> We want to assess whether the student is learning - whether they are becoming an author, whether they produce with autonomy, whether they read and study properly, whether they argue and substantiate properly, etc. - it follows that an intelligent way of assessing them is to assess their "texts". This does not prohibit other forms

of assessment, not even the test (which is generally useless as an instructional assessment), but it does call for preventive and diagnostic virtue as an instrument to guarantee each student their right to learn well (DEMO, 2012, p. 13).

The act of assessment also has repercussions on the attitudes of good teachers and their innovative practices, because even the act of assessing what has been learned requires broader ways of presenting knowledge, not just solving problems that are often the same as the exercises done in class. The teacher must broaden the repertoire so that the student produces and interprets, so that they become the author of their own knowledge. But this also hurts the pride of the teachers who, most of the time, imagine that they are losing their importance, but don't know what they are leaving for the students to evaluate the way we are evaluating them, it's the same as saying that we are teaching wrong, but aren't we wanting to teach more and better?

Couldn't we teach in a way that consolidates knowledge more, so that we can see what the student has really retained?

How do we start innovating or introducing these aspects into education? It is very likely that it will be necessary to change some things in the planning of education at national level, to review the guiding aspects of education, to look for innovative ways, but, initially, it is necessary for the teacher to know how to learn from these innovative practices as well, in other words, a teacher who is able to learn through vast practices will be able to pass them on to their students, which is necessary for the teacher to be able to revolutionize their lessons in a positive way.

One issue that is also much discussed by everyone is the technologies that attract students more easily and make them more interested in their studies. What we can learn from this is that the way in which technology is shown to us is brilliant for exploring this area. We can make the challenge of teaching based on the way in which these technologies are presented to us, starting from them and proposing that students learn by trying to interconnect activities with these technologies, but for all this to happen we need a teacher who knows how to understand all this complexity.

The teacher's daily objective is to build, together with the student, an individual capable of producing their own knowledge, because in life what is most demanded is the ability to construct learning. The teacher, in the current context of education, tries to start from what is on the agenda today in order to build knowledge on top of it. We can also see this attitude as a form of innovation, but the word "innovative" itself carries a great deal of weight. Often there is no point in looking for a new way of teaching, but rather to look at those that were once considered innovative and reconstitute them in a more up-to-date way.

Thinking up innovative ways should not only be the teacher's duty, but a process of getting to know better what we are learning, and for the teacher what we are teaching. We need a revolution, but not a war. The correct term, according to Demo (2012), would be perfection, because perfection is linked to innovation and to everything we have already built ourselves.

Innovate in the sense of preparing, doing, looking for a more appropriate way, because the mentality of our students is more open, if it is something very striking for them it will certainly be significant and soon the objective of learning will become effective, because it will build a sequence of new learning, seek with the teaching method, keep the student focused, dedicated, and most importantly, motivated to achieve the objectives.

When we talk about pedagogical innovation, there are those who already admit to being afraid, because change requires attitude, and attitude requires effort, and effort generates disruption, it's like an addictive cycle, most of the time an innovation isn't started because of a lack of attitude. Many people have the ability to initiate a turnaround, but fear takes over. At this point, as Cunha (1989) tells us, teachers who are the difference have differentiated methods, more specifically in the area of mathematics, we can have various didactic materials, the good teacher must also disconnect from failed methods that no longer add up to students, increasing, pedagogically designed workshops in teaching mathematics.

Working on the content in a different way allows students to think in an extremely different way, much more comprehensively than the old method of writing on the board and the teacher explaining it to the students. In the workshops, the teacher can work with the concept so that they construct knowledge in a more creative way. This way of teaching will be in serious competition with the desire to stay on social networks, chats, chat rooms and others that take away concentration, in this case they are a healthy adversary for the student, something that leaves them confused between good things about school or almost useless and anorexic to the growth of the student, who is also around them.

The teacher's task of thinking of new materials may be a tiresome one, as there are now thousands of websites, as well as blogs with all kinds of materials, from great to bad, all that is needed is a careful evaluation by the teacher beforehand, to select what can or cannot be used in the classroom, as Cunha (2008) proposes that there should be a reorganization of the theory/practice relationship, reviewing these concepts and working minimally with the rupture of paradigms and outdated traditional methods, the possibility of a greater understanding of the content.

The act of reviewing the relationship between practice and teaching, which are not being considered as part of the same plan. At this point it's time to go back and reflect on the beginnings and report that the relationship between theory and practice emerged in the opposite way to the one applied today, because theories were created from facts that happen or are assumed to happen in this or that way.

In the rupturing sense, the reorganization of this relationship assumes that epistemological doubt is what gives meaning to theory. And it arises from the reading of reality. Social practice is therefore the condition for problematizing the knowledge that students need to produce. From this perspective, practice does not mean the application and confirmation of theory, but is its source (CUNHA, 2008, p. 26).

In addition to didactic transposition, there must be mediation involving the parts of the pedagogical process: students, teacher and knowledge.

Mediation is another important category in the paradigm shift, assuming the inclusion of socio-affective relationships as a condition for meaningful learning. It includes the ability to deal with the subjectivities of those involved, articulating this dimension with knowledge. It presupposes respectful relationships between teacher and students, the dimension of the pleasure of learning, a taste for the subject matter and enthusiasm for the planned tasks. Mediation bridges the gap between the affective world and the world of knowledge, including the meanings attributed to it by each individual. Finally, an important condition for meaningful learning. It is a condition for innovation because it breaks with the subject-object relationship historically proposed by modernity. It recognizes that both students and teachers are subjects of pedagogical practice and, even in different positions, act as active subjects of their learning. It includes the participation of students in pedagogical decisions, valuing the personal, original and creative production of students, stimulating more complex and non-repetitive intellectual processes (CUNHA, 2008, p. 26).

Much has yet to evolve in the way knowledge is constructed, involving teaching practices and methodologies. For pedagogical innovation to take place, especially in mathematics, most of the time the teacher uses problem situations.

It is worth noting that in the vast majority of the cases studied, the teachers' moves towards innovative possibilities originate in problem situations, i.e. they start from some discomfort experienced by the teachers in dealing with knowledge or the success of their students' learning. This finding is supported by the thesis of Lucarelli (2003), who argues that difficulties can be generated in any component of the didactic situation; if these are evident as the center of problems, in solving them the teacher develops actions that modify the system of relationships existing between these components, giving rise to learning. (CUNHA, 2008, p. 26)

Another important point, which also aims to give meaning to innovation, is to understand the idea of a problem situation, where, when and how it can be used in the classroom. According to Macedo (2002), we must get to know the adolescent to whom we are proposing problem situations so that we can formulate them in an interesting way, with language that is more appealing to the ears of those who already have many other options to keep their attention. This problem must deal with aspects and situations that we hope to create in a person, in other words, create a momentary expectation so that the student enters a higher level of concentration. Solving problems helps to develop hypothetical-deductive actions.

Perrenoud (2000) describes the practice of teaching through problem situations in ten items. He proposes that these problems should be designed around a situation that the vast majority of students in the class know about and are interested in. The situation should be designed in such a way as to make the student look to their previous learning for tools to solve the problem.

Rays (1991) tells us about the current superiority of the instrumental-technical teaching dimension, forgetting all the epistemological questions of how to construct knowledge, and that it is not enough just to produce it. According to Rays (1991), we are currently treating method as if it were a mere word, without incorporating its conceptions; there is a lack of principles in the definition and in the search for new, more concrete ways to find a more substantial methodology.

The methodology must be thought out in a more concrete way with a view to more elaborate learning, in the sense of approaching knowledge and science in a more coherent and organized way.

The teaching method thus becomes one of the possible elements for structuring the paths to be taken by the didactic action. These paths will use different teaching procedures in order to motivate and guide the student towards assimilating the knowledge conveyed in the school process and its relationship with the environment: natural, cultural, socio-economic, etc. (RAYS, 1981, p. 98).

Methods can be part of the structure and innovation, they are one of the elements of a greater appreciation of didactics in teaching, emphasizing that an adequate teaching methodology is also part of the new innovative practices.

Innovation in the school system was a proposal discussed by UNESCO, and addressed in the book *Teacher Training for the 21st Century* (PERRENOUD, 2002). Thurler (2002), in an article in the aforementioned book, talks about innovative elements that, in addition to teachers, schools should also be responsible for developing. These elements are based on the restructuring of schools that are on the verge of chaos, with the involvement of everyone, with the following aspects being priorities for the configuration of an innovative school:

- They develop common views on how their students learn;
- engage in collective action to put these common points of view into practice;
- collectively take responsibility for their students' progress;
- come together to involve students in the development process;
- develop a collective competence of cooperation that allows them to transform, day by day, the coherence and effectiveness of the learning devices offered by the students;
- are able to gain recognition and thus more support from the school environment (parents, companies, school and political authorities, etc.);
- are evaluated on the basis of a set of indicators: the usual performance indicators, but also student engagement, the individual and collective professional skills of teachers, the pace and effectiveness of the

implementation of establishment projects (THURLER, 2002, p. 92).

When we talk about innovation, our education system even offers some resources for teachers to have continuous training, to be able to attend some kind of course, most of the time offered by the public education systems themselves, but these courses are not valued as they should be by those responsible for education, in other words, for Thurler (2002), even though they are an innovative element for teachers, the lack of value placed on training courses is an attack on the education system itself.

One of the lines of innovation put forward by Thurler (2002) is that of mutual cooperation, as mentioned above. This could be a new way of teaching, through cooperation between students and others involved in the school community. In this teaching practice, the students would learn from what the teacher would send to the class and at this point they would use this practice to learn in a way that they themselves could reach knowledge, but always aiming for the participation of the whole group, in a collaborative way to overcome the difficulties of each one.

The search for the development of a teacher who is capable of articulating these values, emphasizing collective work over individualistic perspectives, and putting into action the new competencies that the new world of education will begin to demand of teachers, the project of presenting the teacher not as a transmitter, but as an advisor, where the real relationship between teacher and student must be based on the articulation of doubts that the system of cooperation cannot handle.

> [...] a new type of professionalism is emerging, characterized by the evolution of teachers' values and practices in favour of a closer relationship between professional development and institutional development. The most significant transformations in teachers' professional culture are the following: the culture of individualism gives way to cooperation; hierarchical relationships are replaced by teamwork; supervision evolves into *mentoring*; refresher courses recede in the face of the popularity of professional development; finally, the contractual approach negotiated between partners replaces authoritarian decisions. (WOODS, et all, 1997, p. 158 apud THURLER, 2002, p. 98).

Perrenoud (2000) states that the first competence of a teacher today is to organize and direct learning situations, but, as Macedo (2002) points out, we must stop thinking that the teaching plan is an anchor from which we cannot detach ourselves; on the contrary, if the teacher feels it necessary, he or she should -invent" his or her lessons, be different with every speech and every action, not become predictable, not let the students fall into monotony, creating the desire to learn is a teacher's weapon.

Teachers who stand out from the rest will always be showing their students the world from a different angle. Machado (2002) points out that students are expected to develop the ability to express themselves in different languages, to understand physical, natural and social phenomena, to make convincing arguments and to be able to project, but for students to be able to display these skills, it would be appropriate for the teacher to be their reference point.

When it comes to innovation, we need to study the subject in more depth and Garcia (1995) shows us very well why there are discrepancies between what is prepared and what is implemented in our education, who want to maintain the old system, where many ideas are based on education of domination, which preaches class divisions, where the most powerful minorities continue to dominate the educationally disadvantaged majorities, and also in opportunities and neglect of society itself.

"[...] the procedures by which these peoples, lagging behind in history, are compulsorily engaged in more technologically evolved systems, with the loss of their autonomy or even their destruction as an ethic entity" (GARCIA, 1995, p. 224).

The evolution of education has never been disconnected from those who brought it from Europe to Brazil. We still live in a dominance of the quality of education, it is a historical sediment of the old and flawed colonial education, where we follow a model proposed by the Europeans for the colonies. Innovation has often been proposed for education and even implemented, but it always falls into routine. It has been proposed a lot since the 1930s, based on different economic models, and every time there is a change, education will be adjusted.

Educational innovation is seen by Garcia (1995) as a change that is judged to be qualitatively better in relation to something that already exists, it means drawing up proposals that aim to break down the barriers that have been created so that we are dominated. For Garcia, the first step is to break down this conservative view of education, changing mainly through the ideals of educators, expressing outside ideas and giving a voice to those who do and live education.

Our teaching methods shouldn't alienate students, they should make them critical, but there is always a need to change education and implement new practices, but these are always seen as something difficult to get out of the rut and change these outdated teaching practices.

For Hernandez (2000), learning means introducing innovation from the very beginning of planning, presenting ideas to teachers, showing them so that they become familiar with and understand what changes in teaching practice, we will see what they express, so that if necessary we can make changes, for a better introduction of innovation into the school system.

> In general, the teachers at the school consider that innovation implies a conceptual change in teaching practice itself, which presupposes constant self-analysis and makes the teacher aware of the learning process that the student is carrying out. For one teacher in the initial cycle, 'working on projects presupposed, above all, having a different idea of what it is to learn and what it is to teach. It may be that in a while, instead of doing projects, we do other things [...] (HERNANDEZ, 2000, p 179)

It is also proposed that innovative higher education training be imposed as a permanent training model, i.e. the prospect of improving the education, and certainly the professionalization, of our students would be the development of strategies involving innovation as a tool for interpreting reality, as described by a teacher quoted by Hernandez (2000) that if innovation does not take into account the reality of the school and the teachers, there is no point.

There is also the idea of innovation in relation to the vision of the future, in which it is suggested that projects be incorporated into school practices. An innovation project only succeeds when it arises or starts from a teacher, and the idea is implemented and a group appears to promote the idea, of which there must be sufficient will to make changes to existing conceptions and maintain the attitude of changing not only the organization of the curriculum, but also implementing these ideals. Maintaining a link with the family and the expectations that the school creates in the minds of the students, thus creating personal satisfaction in both the students and the teachers.

It is at this pace that changes are introduced into the school, and that we accept the contradictions, deal with them and understand them in order to play a better role as a teacher and, most importantly, the need for critical reflection so that it doesn't become a routine practice, as Garcia (1995) had previously told us. We can see that innovation is a sequence of constructions, of projects, of planning, of carrying out research based on themes that the students themselves can come up with.

The idea of innovation associated with change led us to question, when we collected the voices and experiences of others, to what extent these innovations did or did not involve change, and whether they meant an improvement in the school institution. We can see that not all innovations produce change and that change can be positive and/or regressive. The latter are not usually taken into account when talking about innovation, since it is always assumed that it must have a sense of progress and improvement. (HERNANDEZ, 2000, p. 297) .

According to Hernandez (2000), innovation is a process that cannot be implemented overnight. Innovation is a project within a trajectory where goals have been set and plans drawn up, where those involved must be fully aware of the shadows and obscure paths they will encounter. When the concept of innovation is implemented without any basic structure, it is simply an idea that is born compromised, and society itself will not accept these new changes, because as has already been said, to innovate is to change, and change generates discomfort, and discomfort generates work, which many tend to avoid within the context of education.

Many ideas start to diverge within the innovation when the first changes begin. Often, innovation is proposed as a reform of schools, which may also be hindering the implementation of these new ideas. Many schools work with similar ideas, which also need to be articulated, as Hernandez (2000) tells us.

This articulating axis is clearly expressed by the school's first director when he says: - the idea we had of what comprehensiveness and integration meant was an attempt to prevent the disintegration we perceived in our previous experience from occurring. We formulated this idea by saying that all pupils, physically and materially, should be treated equally [...]. For us, all pupils are equal; therefore, we need to offer them material, strategies and subjects that they can adapt and adjust to, but without the more negative aspects of the differences being seen and all of this being experienced as a negative element by the pupils". (HERNANDEZ, 2000, p. 92)

We wanted the students to learn better, from some different content that was closer to their needs [...], some different content that we were used to teaching at BUP and FP, which included everyday facts, current affairs.

It was very lively, and I think it was one of the most interesting things about this process. (HERNANDEZ, 2000, p. 93)

We can see from the principal's statements and from the analyses made for the innovation implementation project that if the important points are really asked for, using strategies that really aim to improve school performance, the implementation of the project will certainly be a success. The intention is to tackle eye-catching themes that address important issues in student, professional and private life, and there must be a bridge between three super important points, which is the implementation of the project, being linked to the second point, which is the technologies, and then reaching the last point, which is the students, to make this action a reality, and thus concretely effect the changes that suggest the new times.

CHAPTER 4: METHODS AND PROCEDURES

Research is one of the most important actions performed by human beings, as it is undoubtedly the greatest source of production, reconstruction and development of new knowledge that makes it possible to understand and intervene in society.

We carried out an explanatory study into the practices of mathematics teachers, looking for significant relationships in order to identify the innovation present in the work of the "good teacher", really identifying what this good teacher is like, discovering how he works and wins over the students.

As a delimitation of the universe surveyed, we opted for students and teachers from schools in the municipality of Pinhalzinho - SC because it is the place of residence of the academic researcher, a total of 11 teachers, six of whom responded to the survey.

The data was initially collected using a questionnaire (Appendix A) applied to classes in the 8th grade of elementary school (corresponding to the 9th grade of the new elementary school system that is gradually being implemented in schools) and the 3rd grade of secondary school ([a]). We believe that the students in the classes at the end of primary and secondary school have the necessary student experience to provide the data for the research.

When we sought out this school/researcher partnership, we took a term of commitment (Appendix C), in which the heads of each institution authorized the application and collection of data, where it was made clear that the real intentions of the project would in no way affect the teachers or their professional ethics and morals. Secondly, we conducted interviews with the teachers from the participating schools. We drew up an interview script (Appendix B) which was presented to the interviewees, some of whom objected to the interview. The aim of the interview was to identify their point of view on their teaching practice, the difficulties they encounter in ensuring that teaching and learning are effective and what the characteristics of a good maths teacher are, according to the maths teachers themselves, and whether what they have been told by the students comes true in the classroom. We have worked on and analyzed the data collected, and we have come to terms with the lack of answers from some of the teachers, pointing out that difficulties will always be encountered.

The students involved in the research were defined on the basis of an analysis carried out by the authors surveyed, most of whom say that the relationship is linked to the minimum time required for students to really get to know their teacher, and thus be able to explain their perception of a good teacher and their respective teaching practices.

With regard to the questionnaires applied to the students, it can be said that there are no incorrect answers in this type of research instrument, as they are the students' proposals, highlighting in their opinion the positive and negative points, according to their reasoning, feelings and certainties. By analyzing the data from the questionnaires, together with the analysis of the teachers' interviews, we will be able to reach a common denominator within the research, reaching its culmination, which is to find the fundamental characteristics of being a good math teacher, indicating the elements that the students pointed out together with what the teachers indicated in their interviews.

The treatment of the data was predominantly qualitative, seeking to find the key issues and to articulate the discourse of the students and teachers with what is theoretically established in relation to this issue:

-Characteristics of a good teacher, especially a math teacher;

-Aspects of pedagogical innovation;

-Use of didactic materials for teaching mathematics.

Let's take a closer look at our field of research, which includes all the schools in Pinhalzinho. There are five schools:

- José Marcolino Eckert Elementary School,
- Vendelino Junges Elementary School,
- José Theobaldo Utizg Municipal School of Basic Education,
- Maria Terezinha Municipal Elementary School (E. M. E. F.) and

- Centro de Ensino Objetivo.

Of these, José Marcolino Eckert and Vendelino Junges are state schools, belonging to the 2ª Gerência Regional de Educaçao (2ª GERED), linked to the State Secretariat for Regional Development (SDR) in Maravilha. They are the two largest schools in Pinhalzinho.

The José Marcolino Eckert primary school is located in the Santo Antônio district, near the parish church, and had 1221 students enrolled in 2013. These students come from various parts of the municipality, from the countryside, the city center and the city districts. Most of them use the school transport provided by the government. There are around 60 teachers at all levels of education, including primary, secondary and teaching. It is the most important school in the city because of its structure, laboratories, equipment, library and ample space. It currently has four math teachers.

In this school, the questionnaire was answered by two classes of 8ª elementary school and two classes of 3rd grade high school.

Table 1. Number of students per class E. E. B. José Marcolino Eckert - 2013.

Elementary School		High School	
Class	N⁰ students	Class	No. of students
8th Series 3	21	3rd Year 3	19
8th Series 4	16	3rd Year 4	25
Total	37	Total	44

Vendelino Junges Elementary School is located in the center of the city, most of the students come from the school's surroundings and also from the interior of the municipality, it has a vast physical space, good laboratories and differentiated materials that count a lot in the students' education. It had 607 students enrolled in 2013. It is the second largest school in the city and caters for the basic levels of education, from primary to secondary, with around 38 teachers, including two maths teachers.

In this school, the questionnaire was answered by a single 8th grade class. The 8th grade 2 class that took part in the survey had 24 students.

We also have the José Theobaldo Utizg School, which is a municipal school located in the Efacip district. It caters for primary education up to the 9th grade, and also works with Youth and Adult Education (EJA) at two levels, primary and secondary. It also has a well-structured physical space. There are 35 teachers, three of whom teach mathematics. The vast majority of the school's 723 students come from the interior of the municipality and nearby neighborhoods.

In this school, the questionnaire was answered by four 8th grade classes, which corresponds to 100% of the students in this grade at the school. We didn't administer the questionnaire to secondary school classes, as they are taught in the school's EJA system, which is different from regular education.

Table 2. Number of students per class E. M. E. B. José Theobaldo Utizg - 2013.

Class	No. of students
8th Series 1	26
8th Series 2	27
8th Series 3	20
8th Series 4	12
Total	85

The Maria Terezinha School is also a municipal school, but only caters for primary education. It's located in the Maria Terezinha neighborhood, serves the vast majority of students from this neighborhood, and is the furthest away from the city center. As it is a poorer and more dangerous neighborhood, the school also suffers from this. It has a more restricted environment with little physical space, and does not have the same infrastructure as other schools. It has one math teacher. It has around 170 students enrolled in 2013 and 14 teachers who have the support they need to work from the management, parents and colleagues.

In this school, the questionnaire was answered by a single class of 8ª elementary school students. The 8th grade class that took part in the survey had 13 students.

The only private school in the municipality, Centro de Ensino Objetivo, also took part in the research. Located next to the Vendelino Junges primary school, it is linked to the Rede Energia de Ensino and works with all stages of education, from primary to secondary, and also teaches children from 3 years old. Most of the students are the children of businessmen or people who have an income above the vast majority of the population.

In this school, the questionnaire was answered by a 3rd year high school class. The 3rd year 1 class that took part in the survey has 11 students.

When the questionnaires were administered, the schools showed interest in the research project and the objectives to be achieved, but even so we didn't get a hundred percent return, because at the time the questionnaires were administered, some of the classes were taking tests, doing work, or in subjects where they weren't allowed to answer the questionnaire, so for various reasons we didn't reach our initial target of one hundred percent.

The research tools were chosen and worked on in the language of the teacher and the students, with questionnaires being applied to the students with open and closed questions, and the teachers being interviewed with questions open to debate.

CHAPTER 5: DATA PRESENTATION AND ANALYSIS

In this chapter of the research report, we will present and analyze the data obtained from the application of the instruments used: questionnaires (Appendix A) answered by the students (8ª grade of elementary school and 3rd year of high school) and interviews (Appendix B) given by mathematics teachers from schools in the municipality of Pinhalzinho - SC, as described above.

Focusing on the perceptions of students in the final years of primary and secondary school and with the aim of surveying the characteristics of mathematics teachers, trying to identify which of them are necessary for a "good teacher", we used a questionnaire as a data collection instrument.

The questionnaire consists of three parts.

The first part allowed for identification of the student's school, gender, age, grade/year of primary or secondary school and class. In this group of questions, the student's name was indicated as "optional" when filling in the questionnaire, in order to give the student complete confidence to answer the questions without fear of being honest in their answers.

The second part involved 8 (eight) open questions that allowed the student to get involved in the research topic by expressing their feelings about mathematics and the special teachers they had in their school life and by giving their opinion on the characteristics of a "good teacher" of mathematics.

The third part of the questionnaire was made up of 30 closed statements (propositions) drawn up by Oliveira to collect data for his Course Conclusion Work carried out in the first semester of 2007, in the Mathematics Degree Course at the Catholic University of Brasilia (UCB). ªWe chose to submit these same statements, with four possible answers, to students from the 8th grade of elementary school and the 3rd year of secondary school in the municipality of Pinhalzinho - SC, due to the proximity of our research topic to that carried out by Oliveira (2007) and also because it made it possible to reinforce/extend the aspects presented by the students in the open questions and, at the same time, compare the information obtained. The closed-ended statements are at the end of the questionnaire (they were printed on the back of the sheet) so as not to influence the initial response to the open-ended questions, but with the richness of multiple aspects to be considered for the characterization of the "good teacher" of Mathematics: in terms of the teacher-student relationship, the teacher's performance in the classroom and teaching methodologies, and assessment practices.

The data obtained from the questionnaire will be presented in two blocks, grouping together the information relating to the perception of the students: (1) from the 8th grade of elementary school in four schools, and (2) from the 3rd year of secondary school in two of the five schools involved in the research. Our intention is not to identify variables linked to each of the schools, so the information obtained in the second and third parts of the questions will be tabulated and analyzed together in a single block.

5.1 Perceptions of students in the 8th grade[a]

We will first present and analyze the information obtained in the first part of the questionnaire, which was answered in the first semester of 2013 by 159 students from eight grade 8 classes[a] from four elementary schools in the city of Pinhalzinho - SC.

Graph 1 Age of students in grade 8[a] by school - 2013.

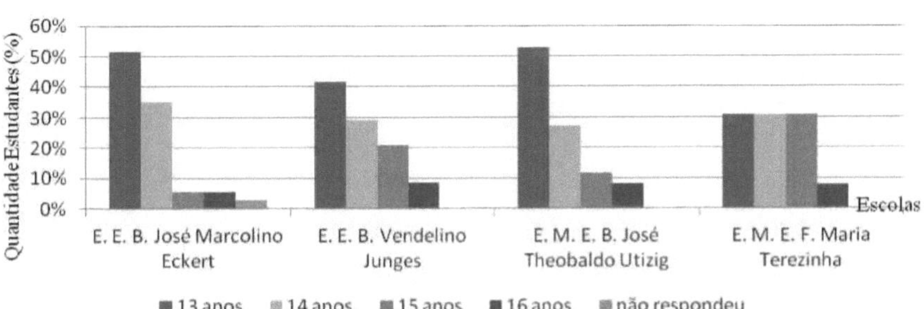

The age of these students ranges from 13 to 16, with 78.7% of them aged 13 and 14. This indicates that they entered the first grade of elementary school at the age of 6/7 and continued their schooling without failing grades or any other type of interruption, which must have been experienced by a smaller proportion of the students (approximately 21%) who are attending the 8th grade at the age of 15 or 16.

Graph 2 Made Students 8ª sene - 2013.

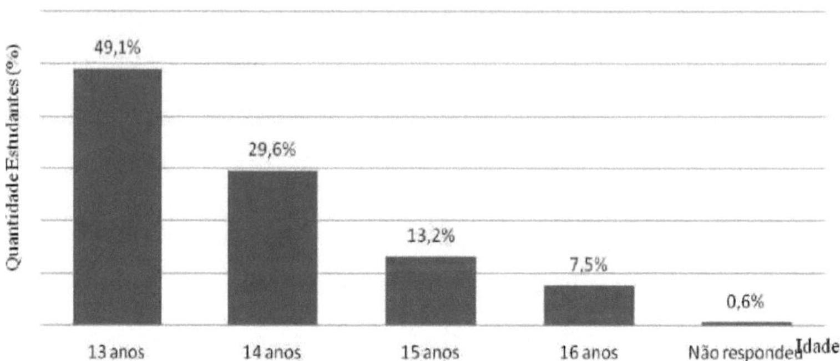

The predominance of one sex or the other in the composition of students in the 8ª grade classes in each school is striking. At Vendelino Junges, the majority (66.7%) of students are female, while at the other schools there are more male students (Graph 3). We can see that when we put the classes from all the schools together (Graph 4), it is male students who predominate, with a difference of 18.2%.

Graph 3. Sex of 8th grade students by school - 2013.

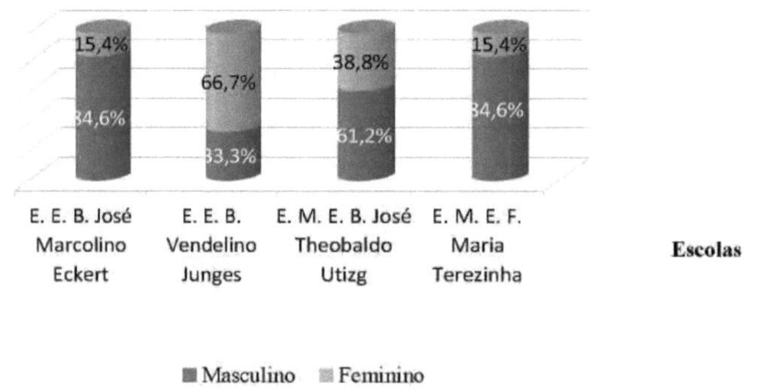

Graph *4*. Gender Students 8ª grade - 2013.

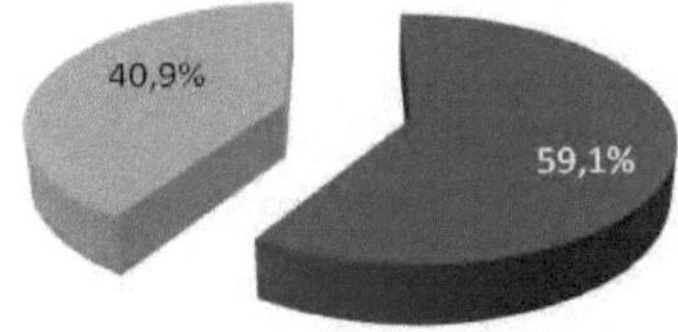

[a]We also noticed that, even though it was optional to identify oneself by giving the name of the person who was answering the questionnaire, very few students in each grade 8 class did not identify themselves, around 21%, which indicates that they were not afraid to answer the proposed questions truthfully.

In the second part of the questionnaire used for data collection, the first question was designed to identify whether the student had had a special teacher who had made a difference to their life. 68% answered "yes" and the other 32% answered "no"

Table 3 shows in alphabetical order the subjects taught by this special teacher and the frequency with which they were mentioned (second question). In a few cases, the same student indicated more than one teacher and subject.

Table 3. Subjects taught by a teacher considered special by students in the 8th grade of elementary school - 2013.

Discipline	Frequency (%)
Arts	5,3
Science	17,9
Physical Education	5,3
Geography	4,5
History	17,0
English	2,7
Mathematics	29,5
Music	0,9
Portuguese	8,0
Religion	0,9
Various subjects (early years)	7,1
Did not indicate discipline	0,9

Total	100

The percentage of students who indicate that the special teacher works in subjects that are part of the curriculum in the final years of elementary school is significant, with only 7.1% recalling as special teachers who worked in several subjects in the initial years of their schooling.

It is interesting that the most frequent indication was for Mathematics. From what was presented in response to the next question (question 3), only one statement indicates that the teacher made a negative impression: "because he asked me what my disability was (I don't have one), then the whole class made fun of me and that changed my life".

The answers mentioned both the personal relationship between teacher and student and the teacher's performance in the classroom and the teaching methodologies used. There was no mention of assessment practices. There was a predominance (68.8 %) of statements that are preferably linked to aspects of the personal relationship between the teacher and the student. Here are some of them.

"Because with her I learned not to listen to what other people say, but to listen to the people who like me."

"Because he was my only math teacher in elementary school, and also because he always treated me very well, and he explains things very well too."

"Because he always helped me when I needed it and helped me learn more."

-She helps me with everything, she's much more than just a teacher, she's a friend." "Because she's the only teacher who understands me."

-She was a teacher who applied companionship, she was a friend to all the students."

-Because she's cheerful, understands the students, likes the same things as us, the lessons are great, relaxed,..."

-Because she was and is very special to me, she helps with both my academic and personal problems, she gives advice, she's friendly and sweet."

-Because she explains everything with love, she's calm and a good teacher."

It is also clear that the teacher's relationship of partnership with the students goes beyond the classroom and the school, as is evident when one student mentions: -helped me a lot in and out of class, me and my family", and others indicate: -well, until then I was a very bad student and she taught me how to be 'people' so, for that reason, I think she was special in my life"; -he taught me to respect others, to dedicate myself more to my studies, that's why I admire him to this day even though I don't study with him anymore". Statements like these reveal the teacher's commitment to an integral education for citizenship.

31.2% of the students' answers emphasized the teacher's performance in the classroom, highlighted by his interest in explaining the content and the attention he gave to students who had doubts, as is clear when it is indicated that -he noticed that the students weren't understanding and explained again", -in her classes I understood and I wasn't afraid to ask her things, I think she's the best teacher, she's calm when explaining", or the teaching methodologies he used: -he dedicated himself to giving different activities in class", -he didn't copy, just because he gave different lessons, he talked a lot, I think we don't just need to copy, sometimes we learn a lot more by talking".

Among the factors that differentiate a good teacher from other teachers, approximately 64% of those indicated were aspects related to the teacher's performance in the classroom and their teaching methodology. What is most emphasized is that the teacher must know how to explain the content in such a way that the student understands it. In this sense, we can highlight a few statements:

-Knowing how to explain and making everyone understand the content."

-The good teacher explains better and has more willpower to explain the subject, not the one who talks, talks and we don't understand anything."

-Knowing how to explain and making everyone understand the content."

-A good teacher explains things calmly and gives students a better understanding."

"A lot of things differentiate a good teacher from others: the subject he explains, the students learn much better than with other teachers." -He explains better and explains until everyone understands."

-The one who explains well, not just the one who explains and tells you to do the exercises." -The good teacher explains several times, answers all your questions, is understanding, helps you and the others pass on activities and sit down."

-A good teacher is one who cares about the student and his way of teaching, the other simply comes in, gives the content and lets the student fail the exams."

It's clear that the students' desire is for learning to take place. What's the point of teaching if learning doesn't take place? Teaching and learning are different processes, which delegate different responsibilities to teachers and students, but which are highly intertwined. There is no such thing as teaching if learning doesn't take place and this is evidenced by a large number of statements in the questionnaire used to collect data for the research. A good teacher is "someone who, in the midst of the 'mess', keeps calm and finds a way, and who knows how to explain properly how each student needs it".

Approximately 36% of statements relate more to aspects of the relationship between the teacher and the student to differentiate the good teacher from other teachers:

-A good character, teaching, good will and respect."

-The one the student likes, the one who explains."

-Your friendship with the students, scolding when necessary."

-Who respects the student and swears only when necessary."

-A teacher who knows how to understand the students and has the spirit of a young person or teenager."

-A good teacher understands the student."

-Your friendship, sincere, calm."

-The one who is attentive, nice and friendly."

The teaching profession is quite complex, it requires us to have a mastery of the scientific content of the subject in which we are going to work, a mastery of pedagogical techniques and methods, it requires us to organize ourselves with good planning and, in parallel to this, to have a love for teaching. This is highlighted in the students' statements: -a good teacher is one who likes teaching, who likes the students, who knows how to explain the subject well, a dear teacher", "the character, the kindness, the patience, the will, the style of teaching. It's not just going to the front to explain and then giving the test to the student".

In the next question, students had to indicate the best math teacher they had and the grade(s) in which he or she taught them, and they could name more than one. This was done by 81.1% of the students in the 8th grade[a] who took part in the survey, the others didn't indicate anything or answered with: -none", -never had and never will have anything against the teacher, but I never liked the subject and the person who explains it", -no cool or special math teacher I've ever met"

The aim was to recall good experiences with math classes so that they could indicate the characteristics that a good math teacher should have and what a good math class should be like. We won't mention the names of these teachers, but we would like to point out that there was a wide variety of recommendations in each class, and there was a predominance of teachers from the final years of elementary school, which is close to what the student is studying. The recommendations allow us to see that the choices are very personal and that the same teacher can be the best for some and the worst for others. At the same time as we see that it is difficult for the same teacher to please everyone, we realize that the indicators presented for a "good teacher" are valuable.

The personal relationship between the teacher and the student was one of the categories used to group and organize the statements made by 92.3% of the students who answered the next question (question 6) regarding the characteristics that a "good teacher" of Mathematics should have. Aspects related to this category were indicated by 26.2% of the 8th grade students. The aspects mentioned involve friendship, understanding and good humor: -being nice and friendly", -making jokes during class and not taking everything we say literally", "understanding and patient", "not being in a bad mood, always cheerful, a guy who likes what he does"

The other statements, made by 66.1% of the students, related to the teacher's performance in the classroom and the teaching methodologies he or she used. Having -patience, knowing how to explain, knowing how to lead your classes, knowing and knowing the subject well enough to want to teach", -knowing how to explain well, understanding what the students don't know, explaining several times if necessary" and -a good explanation of the content, if you need to come back, explain, come back", indicate that the teacher must know the content he or she is working with (have mastery of the content), must be able to get the students to learn and, to do this, must use different strategies and methods (have mastery of didactics), must involve the students in learning, have the sensitivity to understand and clarify each individual's doubts.

"Teaching math in a more fun and enjoyable way" and "there has to be good discipline, help, games, learning and work" are statements that highlight the importance of diversifying teaching methods. It can be seen that the students don't see many different possibilities for their classes, because traditional teaching still predominates, with a lack of meaning and a focus on repetition and memorization. When they repeatedly say that the teacher must -explain well", they yearn for the teacher to act more effectively with new methods and by using different teaching resources and materials, to be -comprehensive, to enjoy teaching and to explain the subject properly", -I think he must have the patience to explain, he must have knowledge, he must love the subject and he must understand that it is often difficult for students to understand".

Another aspect worth highlighting is the need for teachers to have patience. In some statements it is clear that teachers are losing control and getting into friction with students. Statements such as -knowing how to explain and not being rude", -having patience, knowing how to 'get to know' each student, knowing how to explain even when things are messy and that even if they are angry, they don't go off attacking or saying the wrong thing" and -when he speaks more calmly, not shouting and cxplains the questions calmly" characterize the "good teacher" of Mathematics, as well as being "calm, patient, enjoying what he does, trying to teach" and "having a good mood and explaining calmly" show us that there are conflicts between teachers and students (and vice versa) that are damaging the teaching and learning of Mathematics. We know that many students are not easy to deal with, that they are unmotivated, that they disrupt lessons, but we can't give up, we have to persevere and find ways to achieve the proposed objectives. It will certainly be easier if the whole school community (teachers, administrators, parents, pupils) work together to overcome the difficulties.

None of the students listed aspects related to assessment practices as characteristics that a "good teacher" of mathematics should have.

Similarly, 84.4% of the indications that predominate to characterize a "good lesson" in Mathematics are related to the teacher's performance in the classroom and the teaching methodologies they use. The personal relationship between the teacher and the student stands out in only 9.4% of cases. 6.2% of the students did not give an opinion on the question.

"A good math class is good when everyone pays attention and participates in the lesson", "when the whole class pays attention and the teacher explains so that everyone understands, that's a good class". Attention to the student, patience and respect are what again stand out for personal relationships:

"With patience, in silence listening to the teacher, with respect among the students."

"With the teacher helping me, without swearing or shouting at me."

"Being nice, that everyone in the class respects opinions, activities, etc. Good manners."

-The teacher teaching the students in a good mood, if they don't know, the teacher goes and corrects, helps the student. That way the students will learn more and more every day."

The students admit that silence and collaboration are important for a good lesson: "with a lot of explanation and patience and that everyone collaborates, otherwise it's difficult to learn", "without a mess, everyone paying attention to the teacher", "with well-behaved students and a good teacher", "the students paying attention and learning the subjects and speaking one at a time and other things" are some of the many statements in this regard.

There is an emphasis on more dynamic lessons that involve the student in fun activities:

-With different activities, not just from the book."

-I think there should be some different things, like not just copying and counting, doing different things, etc."

"Quiet, peaceful. Like when we're bored, the teacher cheers us up with a movie, something different, some activity outside the classroom and if there's a 'mess', doesn't get out of control by attacking the students, but knows how to explain it properly anyway."

"It's work in the computer lab, films and commentaries."

"Not a lot of content, but some accounts and others that aren't so easy or difficult, and also some movies whenever possible."

"A different lesson, not just doing math."

"When we do a different lesson and others."

"Fun, with games about the subject." "Different activities, lots of work in pairs and little explanation (my opinion)" "Interesting activities, different lessons."

"Different, unforgettable."

"Develop the subject, explain it well, then do the math, give different lessons, work in groups, get out of the classroom a bit."

"Fun, with games, competitions between classmates so that they could think in groups or individually."

"Studying the content well, and sometimes playing games and different activities."

Student involvement is also mentioned, indicating that it is necessary to encourage students to ask questions, to clarify doubts, to give suggestions: "where everyone asks questions with comments and there is silence", "with good explanations, good exercises, everyone participating in class, etc.", "there should be activities, the teacher should interact more with the class, the teacher also has to question the students, etc.".

We can identify aspects related to assessment practices in a single position: -doing a little math, or even a lot, but until all the students have learned, because if you've done everything correctly, you'd take the test without counting." It's interesting to note that in this line of thought, evaluation would be taking place throughout the process and not just at the time of the test, which wouldn't even be necessary because the teacher would only stop with the "math" when everyone had learned.

Graph 5. Taste for Mathematics students 8ª elementary school - 2013.

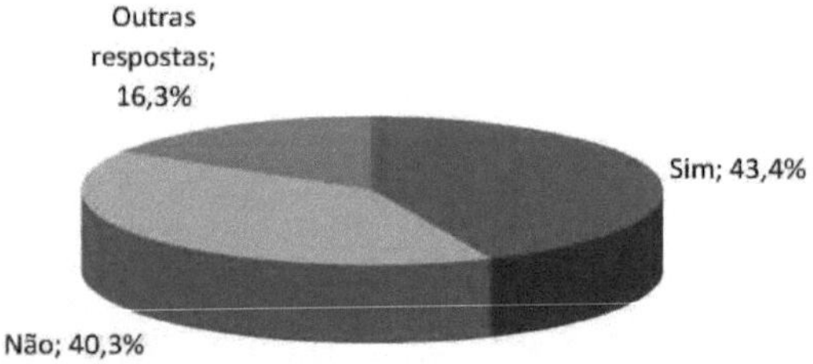

In the 8th grade of elementary school, the indication of liking math (43.4%) is slightly higher than the percentage of those who don't like math (40.3%). The other answers that appeared were: -"more or less" or "sometimes", as can be seen in the statements - "more or less because I like some content and others I don't, like the ones I can't understand", "sometimes because there's too much complicated math or too much content".

When we tabulated the data, we saw that this fact was repeated in much the same way in each of the schools that took part in the survey.

Those who like mathematics can justify it, especially from the start:

- of the importance for life, professional activity and continuing studies; we have 40.6% of statements such as "because it's something I'm going to use all my life", - because it's what we need to learn most in order to work in many professions" or - because I like electronic engineering, because to be an engineer you have to know the math at the tip of your tongue";
- of the ease with which they learn their content. "Because I'm easy to learn, I've always been good at mathematics" is the idea found in 14.5% of the answers;
- of liking mathematics, of feeling challenged in their activities, "I think the word like means love mathematics, I think mathematics is important, yes, to learn to reason and so we have a good evolution", is what the rest of the students say.

Those who don't like math justify it, especially (54.7%) on the basis of the difficulties they have in learning the content. This is evident in -because it takes me a long time to learn it and when I do, the teacher is already teaching something else, so I get left behind", "because I have a lot of difficulties" or "I don't like it very much, I have a lot of difficulty but I make an effort to learn it". Other aspects mentioned, less frequently, refer to the teacher: "because of the teacher", "because teachers often don't explain things properly"; they also refer to not being useful beyond the classroom: "because it teaches few things that I will use in my social life", "because there are certain things that I learn in math class, but I will never use in practice".

Below we present and analyze the results obtained with the third part of the questionnaire used to collect data from the students.

This final part of the questionnaire was presented on the back of the sheet containing the identification data and the open questions. It contained questions for the student to select one of the answer options. In addition to the options "always", "often", "rarely" and "never", which were included in the questionnaire, we have added the options "none" in the tables below, to indicate that none of the options were chosen, and "more than one" to indicate that more than one of the possible options was chosen in the same item.

The propositions presented included multiple aspects to be considered in order to characterize a "good teacher" of mathematics. These are related categories:

(1) the personal relationship between teacher and student,

(2) teacher performance in the classroom and teaching methodologies and

(3) evaluation practices.

Table 1. Perception of students in the 8th grade[a] regarding the personal relationship between teacher and student - 2013.

Proposition	Always (%)	Often (%)	A few times (%)	Never (%)	None (%)	More than one (%)
Teachers who have a love of mathematics are able to make students enjoy learning it.	35,2	40,3	22,0	2,5	-	-
A teacher who doesn't relate well to the students makes them enjoy the class less.	41,5	33,3	13,9	10,7	0,6	-
An attentive and friendly teacher encourages learning.	78,0	19,0	1,8	0,6	0,6	-
Teachers who are enthusiastic about their subject motivate me to study more.	40,3	37,8	18,2	2,5	1,2	-
When you can't learn the content, do you just memorize it?	3,1	22,0	48,4	24,6	1,9	-
The students' opinions are valued by the teacher.	37,7	33,3	23,3	4,4	1,3	
I know I can count on my teacher even outside the classroom.	23,89	29,59	26,41	19,49	0,62	

From the percentages indicated (Table 1), we can see that the predominance of answers that were marked in each of the propositions presented were very close to those obtained in a study carried out by Oliveira (2007).

It was expected that the students would positively value good personal relationships with their teachers. The indications made do not contradict what was presented by the students in the previous part of the questionnaire, which included open-ended questions.

With regard to the last item, we see that "rarely" or "never" is a recourse for students in grade 8ª to memorize content without learning it.

The percentages obtained for each item that could be associated by 8th grade students in relation to the propositions that refer to the teacher's performance in the classroom and teaching methodologies (Table 2) show that students value differentiated activities and that math teachers, or teachers in general, are not very creative in their lessons. They still focus a lot on traditional practices, possibly because of the conditions in which schools currently find themselves, and also because they adopt an attitude that is common to most teachers and is considered normal. The textbook appears to be the main resource used.

Chart 2. Perception of students in the 8th grade of elementary school regarding the teacher's performance in the classroom and teaching methodologies - 2013.

Proposition	Always (%)	Often (%)	A few times (%)	Never (%)	None (%)	More than one (%)
Does the teacher explain how mathematical knowledge has developed throughout history?	18,3	19,5	43,4	14,5	0,6	-
The teacher fully reproduces what is taught in the textbook.	29,6	50,3	16,4	3,1	0,6	-
The teacher teaches mathematical concepts according to the student's reality, presenting examples and problems related to their day-to-day lives.	26,4	39,0	25,8	8,2	0,6	-
The teacher goes through and corrects lists of exercises in order to grasp the content.	45,9	30,2	17,6	4,4	1,3	0,6

Are the students shown various examples and illustrations so that they can better understand the content?	45,3	32,7	18,2	3,1	0,6	-
Can the teacher identify your difficulties and see if you really understand the subject?	36,5	30,8	22,6	9,4	0,6	-
The teacher encourages students to participate in class.	39,0	35,2	22,6	1,9	1,3	-
When a student asks a question that isn't related to the content being taught, the teacher doesn't go into the question and says that this subject will only be studied later or next year.	27,0	27,7	32,1	13,2	-	-
The math teacher is patient when explaining topics that seem difficult to understand.	51,6	28,3	11,3	7,5	1,3	-
The teaching materials provided by the teacher (handouts, books, copies, websites, etc.) help us to understand the subject.	33,3	43,4	15,1	5,7	2,5	-
Math lessons become more interesting when the teacher uses other resources such as videos, games, computers, etc., varying the style of the lesson.	60,4	17,0	11,9	9,4	1,3	-

Chart 2. Perception of students in the 8th grade[a] regarding the teacher's performance in the classroom and teaching methodologies - 2013. (Continued)

Proposition	Always (%)	Often (%)	A few times (%)	Never (%)	None (%)	More than one (%)
It's nice when the teacher teaches "tricks" so that the student remembers how to use the mathematical formulas.	37,7	35,8	15,7	9,4	1,3	-
It's important to check what the student already knows about a particular subject so that the teacher doesn't repeat too much of the same content.	41,5	35,3	15,1	6,9	1,3	-
A math lesson becomes more interesting when the teacher is able to relate it to other	37,1	32,1	18,9	11,4	0,6	-

disciplines.						
Encouraging students to take part in championships, olympiads, lectures or fairs involving the teaching of mathematics helps to awaken their interest in this subject.	34,69	28,3	27,0	9,5	0,6	-
The student's interest in mathematics is stimulated when the teacher develops activities outside the classroom, such as laboratories, sports grounds, visits to other places, etc.	54,7	23,3	15,1	6,3	0,6	-
Much of what I learn seems to have no meaning in my day-to-day life.	20,1	24,6	34,6	18,8	1,9	-
When I can actively participate in the lesson, asking questions and giving examples, I enjoy the lesson more.	48,4	29,5	15,1	6,3	0,6	-

We can conclude, as also indicated by Oliveira (2007), that the teacher manages to meet the desires and needs of some of the students, but not all of them, which is evidenced by the distribution of answers in the different possible items.

There is a very significant percentage of students who support the use of other resources (videos, games, computers), varying the style of the lesson and carrying out activities outside the classroom (in laboratories, sports courts and field trips). This is also true of the last proposition in Table 2, relating to student participation, which had already been mentioned in the open-ended questions as an important point in math classes.

Chart 3. Perception of students in the 8th grade[a] of elementary school regarding evaluation practices - 2013.

Proposition	Always (%)	Often (%)	A few times (%)	Never (%)	None (%)	More than one (%)
The class receives guidance on how to study more efficiently.	34,6	35,2	19,5	6,9	3,8	-

Does your teacher only prepare you to take his exam, or is he concerned with preparing you to transform your life, the environment in which you live and to face the problems of the world?	40,9	27,7	20,8	7,6	3,1	-
The teacher's comments on your performance help you to improve the way you study and learn.	38,4	40,3	13,2	6,9	0,6	0,6
The teacher has a habit of giving a more difficult test than the exercises done in class.	22,6	27,7	33,9	11,9	3,8	-
The teacher adopts alternative assessment methods (individual and group work , reports...), surveys, etc).	27,1	31,4	29,6	11,3	0,6	-

In relation to assessment practices, we can see that the students also distribute their answers across the different items, which again indicates that the teacher manages to meet the expectations of some, but not all.

5.2 Perception of 3rd year high school students

Initially, we will present/analyze the information that characterizes the fifty-five 3rd year high school students from two schools in Pinhalzinho - SC who took part in the research. This data was obtained from the questions proposed in the first part of the questionnaire applied in the first semester of 2013.

We found that 27.3% of students in the 3rd year of secondary school did not identify themselves by name when answering the questionnaire. This number was slightly higher than in the elementary school classes, which we identified as a lack of interest in contributing to the research, as more than one item was left blank in the same questionnaires.

The age of the students ranges from 15 to 19 years old, with a predominance of 16 year olds (50.9%), followed by 17 year olds (29.1%), as can be seen when we put the information together (Graph 6). This indicates that there were no failures or other interruptions during schooling. A smaller proportion of the students (18.2%) are in the third year of secondary school at the age of 15, so they will be able to enter higher education very early.

Table 5 Age of students in the 3rd year of secondary school by school and class - 2013.

	3rd year 3 E. E. B. José Marcolino Eckert (%)	3rd year 4 E. E. B. José Marcolino Eckert (%)	3rd year 1 Centro de Ensino Objetivo (%)
15 years	0	0	18,2
16 years old	57,9	40,0	63,6
17 years old	31,5	32,0	18,2
18 years old	5,3	12,0	0
19 years old	5,3	0	0
Did not inform	0	16,0	0
Total	100,0	100,0	100,0

Graph 6 Age of 3rd year high school students -2013

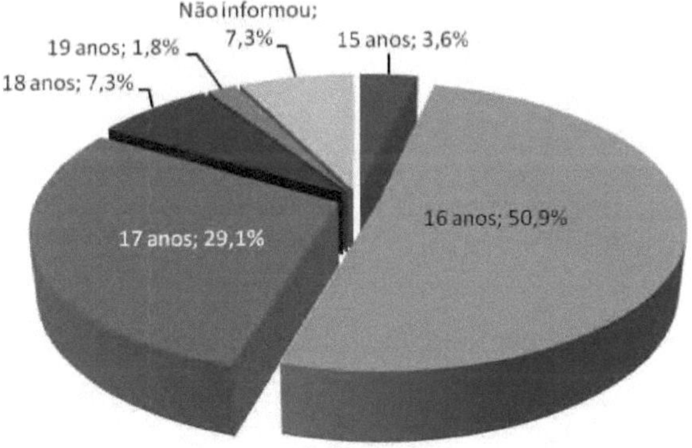

Há the predominance of one sex or the other in the composition of the students in two of the classes and in the other, the number of students of each sex is closer (Graph 7). We can see that when we put the classes together (Graph 8), the percentage of male students (43.6%) is slightly lower than the percentage of female students (52.7%), with a variation of around 9%, with the possibility of an error in this comparison due to those who didn't answer, which could bring the percentages closer together or further apart.

Graph 7. °Sex of students in the 3rd grade por School and Trinas - 2013.

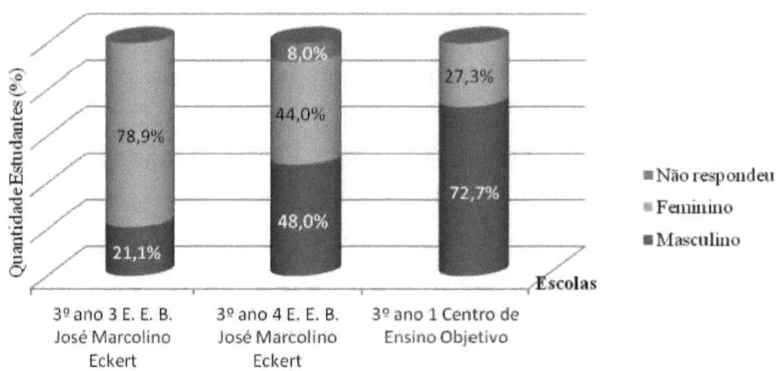

Graph 8. Sex of students in the 3rd year of secondary education - 2013.

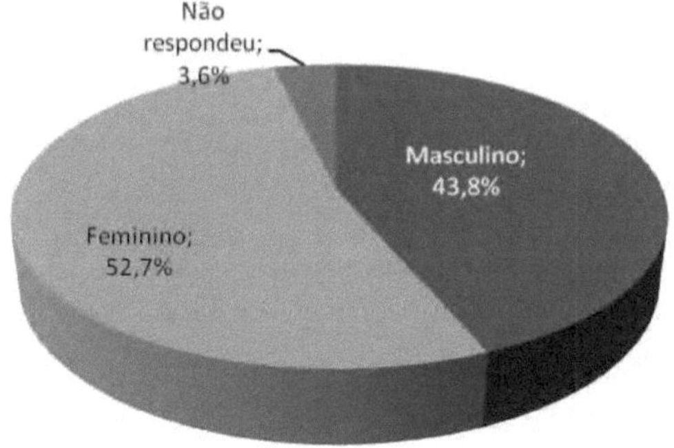

The first question in the second part of the questionnaire sought to identify whether the student had had a special teacher who had made a difference in their life. 9.1% left the question blank, 18.2% answered "no", 1.8% answered "I don't know", 61.8% answered "yes" and 9.1% gave other answers: -*"no,* for me they all made a difference", -*"no,* they all made a difference in some way (good or bad)", -"they all make a difference, because from each one I learned something I didn't know", -"no (for me they were all special, each in their own way)", -"all the teachers are special". We can consider this last group of answers to be affirmative, so we would have an expressive figure of 70.9% of students indicating that they had teachers who were special in their lives. In the same way, we could combine those who left the answer blank with the answer no, obtaining 27.3% of denials, representing a significant but not predominant figure.

Table 6 shows the subjects taught by this special teacher and the frequency with which they were mentioned in response to the second question on the data collection instrument. It should be noted that in some cases the same student indicated more than one teacher/subject.

Table 6. Subjects taught by a teacher considered special by students in the 3rd year of secondary school.

Discipline	Frequency (%)
Arts	7,5
Biology	15,0
Science	7,5
Philosophy	7,5
History	2,5
English	2,5
Literature	2,5
Professional Marketing	2,5
Mathematics	22,5
Portuguese	10,0
Chemistry	12,5
All (first teacher)	5,0
Did not indicate discipline	2,5
Total	100,0

We see that the highest frequency of responses occurred for the subject of Mathematics.

We also tried to identify the reasons why the students considered their teacher to be special (question 3 of the questionnaire). We will use the following categories to organize and analyze the responses, as we did with the data for the 8th grade[a] :

(1) personal relationship between teacher and student,

(2) teacher performance in the classroom and teaching and learning methodologies

(3) evaluation practices.

The answers were analyzed in general, without linking them to one or another subject that this special teacher taught.

Of the third-year high school students who said they had a special teacher (61.8%),

29.4% gave reasons related to aspects of the personal relationship between the teacher and the student.

With indications that the special teacher is able to establish a good relationship with the students, the characteristics highlighted are that of being "nice", "great person", -calm", "outgoing", -always ready to help", as we can see in the following statements made by the students:

-Because he's humorous and nice."

-Because as well as being a great teacher, she was a great person, with great energy."

"She was a calm, outgoing person, and any problem or question she was always ready to help with, she taught the students."

The fact that the teacher values and encourages everyone, taking into account their individualities, also stands out in the statements made:

-Because she always made sure that the morale of any student was high."

-Because I really liked her and she was always able to explain things to me and help me when I had any questions."

"Because she was patient and listened to the students' opinions."

Aspects of the relationship between teachers and students that go beyond the classroom are also covered in the justifications given:

"They were both special and still are, because they teach us to see things differently, both in discipline and in other situations."

"He taught me how to get along better in the most varied situations. He taught me to understand some things in life."

"It's helping me get into the university I've chosen."

Aspects related to the second category used to organize the data collected, which refers to the teacher's performance in the classroom and their teaching methodologies, were predominant (58.8%) in the students' responses.

Statements related to teaching methods stood out:

"Because he taught well, he was fun and he taught well."

"The way she explained the subject made it easy and fun."

"Because the lessons were dynamic and fast, where he could 'dominate' the class."

"Because he knew how to explain the content without boring the students, he was a lot of fun and made the lessons fly by." "Because he teaches in a way that makes the subject more attractive." "I liked the way he explained the content."

"Because he knew how to explain very well, he was dynamic with his lessons, he brought lots of examples, so the lesson didn't get tiresome."

"Because she knows everything you ask her, or she tries to explain it to you. I love her lessons, they're very good, she uses different methods of explanation."

In parallel with the teaching methods chosen by the teacher, there is a clear need for the teacher to have a mastery of the content to be taught, which enables them to make associations with practical situations, to get the students to focus on their studies and not to get lost in side conversations that disrupt the class. In some statements, aspects of the personal relationship between the teacher and the student are mixed in with the teacher's performance.

Learning is also highlighted as a result of the teacher's performance and teaching methods:

"Because I had never understood mathematics in other years."

"Because when we need help, she explains the questions we have very well."

"Because of the way he passed on the content, always in high spirits, always looking for the best way for the students to understand the content."

"Because she taught me how to make notes that will help me in the exam later."

-He was a strict teacher, because he demanded a lot from me and I learned a lot as a result, I ended up paying more attention to what he said and the formulas he gave me."

-Because of his teaching method, he made lessons fun and enjoyable to attend. And I learned a lot from him."

-Because with this teacher I was really able to understand a lot of content."

-Because I like the subject, and when she explains it, everyone understands, the way she gives the subject, it's not just theory."

-Because of the way they taught, their way made me learn well."

Here we can emphasize that the teacher seeks to promote the learning of each student and not just to teach. We are clear that teaching only takes place when learning takes place.

No student referred to assessment practices (category 3) as a factor that justifies the teacher being special and 5.9% did not justify it. The same percentage, 5.9%, indicated that their first teacher was special, that they began to learn with her, which is why it was significant.

As for what differentiates a good teacher from other teachers, 9.1% didn't answer, and the same percentage didn't highlight any particular characteristics, giving statements such as: "I think all teachers are good." and -All teachers are nice, I can't tell the difference." The remainder is divided, with a very close percentage, between aspects of the personal relationship between the teacher and the student (category 1), with 38.2% of the statements, as evidenced by the statement "I think they're all good, but some find it easy to interact with the young style." and aspects of the second category, with 43.6%, relating to the teacher's performance in the classroom and teaching methodologies.

We can highlight the indication that the lesson needs to involve the student. In this sense, two aspects stand out: being associated with the student's enjoyment of learning and the student's participation. With regard to the first aspect, here are a few statements:

-Teachers with funny lessons, with good explanations, make all the difference to learning, you can learn by joking and laughing." -A good teacher is one who manages to win over the class with his way of teaching, he shouldn't be serious, he should be fun and attentive." -The way he explains, acts and treats the student, the way he acts in the classroom." -He knows how to relax the students and make them enjoy his subject, presenting a nice lesson without being boring."

"Friendliness, being fun, knowing how to explain the content well, being very objective."

-The way of being with the students".

"The way we explain things and understand the student's side as well, being the student's friend."

"Who knows how to explain, who is liked, etc. The teacher has to be a friend to the students, then we understand more."

"Good humor in class, interactive lessons."

"Because you can understand better, you enjoy the lessons and you learn better..."

As for the second aspect, we present the students' comments:

"They're all good teachers, one or the other can explain more or talk to the students more, teaching them more.

But they're all the same to me." "A good teacher is one who not only teaches, but also learns from the student."

"His method of explaining the subject, his way of 'making himself understood', his dialog with the class, etc."

"The way he teaches, expresses himself and his knowledge of a particular subject, making the lesson interesting, and the teacher who questions the student a lot."

The teacher's commitment to the teaching process and to the student's effective learning, and not just to teaching his or her classes, identified in the patience with which he or she explains and answers questions, also appears in the students' comments:

"The one who knows how to teach the content, explains it and knows how to work. Not the one who comes to the classroom to hang around."

-Knowing how to calm students down when they don't understand an explanation."

-Patience and politeness/respect."

-Comprehension."

-A teacher who understands the students, has the patience to explain the content of his subject, does different activities."

-In my opinion, a good teacher differs from others in that he doesn't mind explaining things twice, he's the one who helps, who really answers questions."

-The ability to explain and for everyone to understand."

Love for what they do, with regard to their love of mathematics and teaching, were also pointed out by the students as characteristics of a good teacher: -A good teacher stands out from the others when he teaches with love, and has affection for his students."; -A good teacher must have love for what he does, patience, and show interest in the students."

It's also clear from the expressions:

In the statement "Those who help with problems outside the classroom, which often reflects on performance" we can see that the role of a good teacher goes beyond the classroom, and that they need to be sensitive and understand the student as a human being in their entirety.

In the statements: "It's not enough just to have knowledge, but to know how to pass it on.", "Your knowledge and the way you explain the content.", "Liking what you do, and knowing the content clearly." "A good teacher must have great understanding, know how to explain and be humorous." It is clear that mastery of the content they are working with is a basic condition for a good teacher, but it must be linked to teaching methods. The teaching methods are not mentioned directly, but are evidenced in expressions that refer to good lesson planning and how to explain the content:

-There are teachers who come and just want to give content and content, they want to do everything and they don't explain it properly." 'The teaching methods, well-designed lessons..." 'Lesson planning.

"The difference in a good teacher is the explanation."

"The pace of the explanation, the attention and the explanation."

-Dynamic."

'There are no good or bad teachers, there are only those who convey the content better."

-Nothing, only the way of teaching can sometimes be the same."

-Better teaching that isn't always the same concept, that has different lessons." -Explanation, interest and good performance in the classroom."

-They distinguish themselves by being able to teach what they know."

"He knows how to conduct a lesson without making it tiring and he interacts with the students."

In this respect, it was also considered important to direct lessons towards the students' interests: -"A good teacher needs to be patient, know how to articulate and teach tricks for the entrance exam, pass on some of their knowledge and also be open to criticism and change. As they are in the third year of secondary school, preparing for higher education is one of the students' concerns.

The last four open questions in the questionnaire administered to students in the third year of secondary school refer to mathematics: the teacher, the lesson and the love of mathematics.

The fifth question asked students to name the best math teacher they had and the grade(s) in which they were taught. They could name more than one teacher. The aim of this question was to remind them of good experiences in math classes, giving them a basis for indicating the characteristics that a good math teacher should have and what a good math class should be like, since previously, in question 4, the students gave indications for a good teacher considering all the subjects in the school curriculum or the teacher who had made a special mark on their life. We have therefore chosen not to name these teachers.

The categories used to organize the data obtained were the same as those used previously. More than one aspect is mentioned in the same statement, so we chose to quantify it in more than one category.

It is clear that what the student wants is for learning to take place, so he brings up different characteristics that should be present in a "good teacher" of mathematics:

-Knowing how to teach the content in a way that makes it easier to learn, but also how to show other ways. Teaching everyone indifferently, varying lessons, applying different types of assessment and being a friend."

-You must be clear and objective with the content, have a good rapport with the class, explain in a way that everyone understands and answer the students' questions."

"Teaching calmly, not just passing on the content and leaving the students with virtually no understanding of anything, those who are really willing to teach."

The indications presented for a good relationship between teacher and student (category 1) refer to the (good) humor and friendliness that a good math teacher should show, with a frequency of 11.3%, evidenced in expressions such as "you have to be fun, explain the subject well" or "be friendly, good people, charge when you need to". In this same category, 34.0% of the statements emphasize that the teacher must be patient and understanding of the students' difficulties: -speaking slowly, explaining well and being patient, making the lesson more dynamic so that it doesn't become a monotonous, boring routine"; "polite, understanding, dedicated to what they do, patient when teaching"; "knowing how to deal with the class, going at their speed and not his".

In relation to the teacher's performance in the classroom and teaching methodologies (category 2), what was highlighted with the greatest emphasis (49.1%) is related to the teacher's explanations, which must be clear so that the students understand the content, as we can see in the following statements.

-Knowing how to explain formulas well, in detail, how to do the math, teaching everything that is necessary in a way that the students will understand."

-Knowing how to explain in a way that we students can understand well."

-Knowing how to explain very well what you want to pass on to the student ..."

-Knowing how to teach a class, having good knowledge, knowing how to explain the content and how to charge the student as well."

-Being able to explain the content, being able to explain it as many times as necessary."

-explaining and teaching the student well."

-You must explain well, provide assistance if there are any questions."

Students don't always indicate what the process of "explaining well" would be like, but 13.2% refer more directly to teaching methods.

-Trying to interact with the students in the best possible way, always bringing new things (at least one difference, so that it doesn't get boring)." -Don't do the same thing every lesson, teach other things."

-Having good teaching methods, such as explaining the content calmly, with lots of examples..."

-To know the subject and use new techniques that are easy to learn."

Other important but less frequently presented aspects are worth mentioning.

Just over 11% refer to mastery of the content they are going to teach and good planning: -mastering the subject and having patience"; "someone who shows an interest in teaching, and who knows the content they are going to teach"; "knowing the content without having to check it all the time, having planned and objective lessons, but knowing how to act when there are unforeseen circumstances".

15.1% indicated the need to mobilize student involvement and interaction in mathematics classes, including the need for the teacher to consider all students, and to consider them in the same way, which can be seen in the statement - be patient, know how to explain and don't pick students to adore! Treat everyone equally".

-[...] applying different types of assessment" is an aspect related to category 3, which involves assessment practices, and was mentioned by one of the students as a component of the characteristics that a "good teacher" of Mathematics should have. Two students did not answer the question.

The characteristics of a -good class" in Mathematics (question 7), according to the 3rd year students, refer less to aspects of the personal relationship between the teacher and the student (7.3%), with the indication that it should be -a good mathematics class would be a class that is closer to the students!" and that there should be -a teacher who relates well to the students and has good communication skills" and assessment practices (3.6%), being -with consultation tests and pair work" and -not having too many assignments, tests".

The remaining percentage of those who answered this question (approximately 76%, as 12.7% left the question blank) pointed to aspects relating to the teacher's performance in the classroom and teaching methodology. In this category we could group the statements according to different aspects. Pleasant, enjoyable class:

"The kind where there's time for relaxation and time for learning."

-Fun."

-Learning by laughing and playing."

"A productive and enjoyable class."

-A fun, explanatory lesson in which the teacher follows the students' development and makes them understand the subject." -Fun, interesting."

-With fewer bills and more energy because it's always the same. More fun classes."

A class that seeks to facilitate learning with the help of the teacher:

-Fun, with surprising learning, and with a teacher who knew how to explain patiently and see each person's difficulties in order to help those who had more difficulties."

-With explanations in the students' language."

-Less complicated."

"Being well explained and not wanting to do several activities and not explaining them well." "Activities with the teacher's help."

-With individual help from the teacher to better understand the subject and activities."

Differentiated lessons using different resources and strategies:

-A dynamic lesson without being boring as usual."

-With different activities, but within the content. A good explanation also makes the lesson better."

-With different things, perhaps outside the classroom. Not just calculations."

-Having a difference, not being those classes like exercises, corrections and exams."

-Making classes different."

-Differentiated."

-A good math lesson for me is one that varies from time to time, that teaches and that also brings something new to the student."

"Content that is interactive, and not those 'chair-sitting' classes where you can't talk or look around."

-With different dynamics, with beads. I learned in a course when the teacher brought challenges to every lesson, that was very interesting." -Communicative, discursive."

-Full of activities and dynamics."

-Good explanations. Activities. Jokes, games ..."

-There could be different things to stimulate the students more, but the 'normal' content (taught in class) is necessary."

-With various exercises and dynamics."

We can point out that high school students are *already* much more critical than those in elementary school, and their suggestions are more directed towards constant innovation in the classroom.

There are also comments describing the math class we've been used to for a long time:

-Where the teacher explains and then gives some example questions, after which the student must show that they have understood the explanation by doing the questions themselves. And there must be cooperation between student and teacher."

-With the teacher being able to go over the content, asking and solving questions, clearing up doubts."

-Lots of calculations and plenty of exercises for us to learn the content well."

Graph 9. Fondness for mathematics among 3rd year high school students -2013.

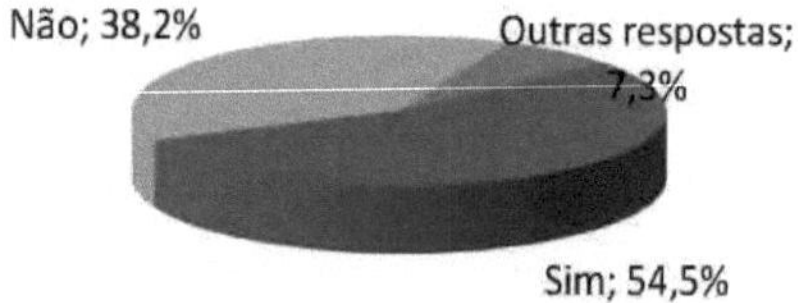

In terms of liking mathematics, there was a predominance of positive ratings, which shows a good improvement on what we had identified in elementary school (students in grade 8ª).

What is indicated as other answers refer to -more or less" and -a little, being justified that -it depends a lot on the content", or emphasizing its importance: -because it will help you ahead, in college, and math is in everything, in most courses" or that -sometimes it's very difficult and complicated to understand, so it makes classes boring".

Most of the reasons for liking or not liking mathematics are related to whether or not they achieve good learning results:

-I always liked the subject. Because I've always been able to understand it and do very well in my grades."

-I love mathematics because I grasp the content easily."

-When I understand, I really like it.

-I'm good at numbers."

-Because it's a subject that I can learn easily because it appeals to me."

"Because I like accounts, I do 'well' in the subject."

-Because I'm doing well.

"It's a legal matter when there is understanding."

-Because I'm more or less."

"It depends on the content, because I don't do well in all subjects, each person has skills and is better at a certain subject, but math is good."

"Because I'm not good at exact sciences."

"Because I'm bad with numbers."

"I don't like numbers very much."

"I have a bit of difficulty."

"Because I've never liked finding y, x, ... I can't explain it, I just know that I've never 'got on' very well."

"I don't find the subject easy."

"Because I hate accounts, I can't get them right."

-It's difficult.

-Because I can't concentrate to do the calculations."

-I have some difficulty.

Another aspect considered was the applicability of mathematics, its importance in solving everyday situations, as we can see in the following statements:

-Over time I've come to love it and I've seen that I can use a lot of what I've learned in my day-to-day life."

-"Because without mathematics we're nothing nowadays, everywhere we go, or any purchase we make, it's involved."

-Because bills, like it or not, are part of our daily lives."

-I don't like it, but it's very important for our whole lives."

-Because I use it every day, even if liking it doesn't mean loving it, I admire mathematics even if I find it difficult."

Interest in mathematics was also associated with skills in the exact sciences - "my area is the exact sciences" - and the direction of the higher education course he wants to take: "because I want to do mechanical engineering, which involves a lot of mathematics", "I like calculations and I want to do mechanical engineering".

Another aspect that deserves to be highlighted is the love of mathematics for pleasure, for developing reasoning and problem-solving skills, which was cultivated from an early age:

-Because math is cool and involves a lot of things."

Because it develops reasoning and problem-solving skills."

-Because I've always identified more with this subject than with others and because I like doing math."

-Because I love doing math, vacations seem like torture."

"Because I like numbers, calculations and everything that involves them, ever since I was little."

-It challenges my ability."

-Because I like doing math and when I don't get the right result I always try to go back and get it right."

-I'm passionate about numbers and the way they're applied, I love to think, and mathematics requires a lot of that."

Or, in the opposite way, associated with displeasure with mathematics:

"It's a lot of calculation."

"Because it's practically a calculus-only subject and I don't like calculus."

-Because I don't have the patience for long accounts. I prefer a book, which helps me with various subjects and which I'll use more in my day-to-day life."

-Because it's no fun."

There were also students who did not justify their choice. There were 1.8% of those who said they liked Mathematics, 10.9% of negative answers and 1.8% who said they liked it more or less, for a total of 14.5%.

Next, we present the results obtained from the third part of the questionnaire used to collect data from the students.

As with the students in grade 8ª , this final part was presented on the back of the questionnaire sheet. It contained propositions for the student to mark one of the answer options.

In the tables below, we have added the options "none", to indicate that none of the options were chosen, and "more than one" to indicate that more than one of the possible options was ticked in the same statement item presented.

The propositions for the characterization of the "good teacher" of Mathematics were grouped according to the following categories:

(1) the personal relationship between teacher and student,

(2) regarding the teacher's performance in the classroom and teaching methodologies and

(3) regarding evaluation practices.

Table 4. Perception of students in the 3rd year of secondary school regarding the personal relationship between teacher and student - 2013.

Proposition	Always (%)	Often (%)	A few times (%)	Never (%)	None (%)	More than one (%)
Teachers who have a love of mathematics are able to make students enjoy learning it.	21,8	60,0	12,8	3,6	1,8	-
A teacher who doesn't relate well to the students makes them enjoy the class less.	72,7	16,4	9,1	1,8	-	-
An attentive and friendly teacher encourages learning.	67,3	29,1	1,8	-	1,81	-

Teachers who are enthusiastic about their subject motivate me to study more.	30,9	49,1	14,6	-	3,6	1,8
When you can't learn the content, do you just memorize it?	5,5	34,6	40,0	18,2	-	1,8
The students' opinions are valued by the teacher.	25,5	40,0	29,1	5,5	-	-
I know I can count on my teacher even outside the classroom.	9,1	34,5	41,8	10,9	1,8	1,8

If we compare this with the data obtained from students in the eighth grade, we see a clear similarity: high school students also indicated that the personal aspect between teacher and student also counts when it comes to thinking about teaching and learning.

A very well put point is that most motivated teachers motivate their students too. Another good sign is that teachers value their students' opinions.

Chart 5. Perception of students in the 3rd year of secondary school regarding the teacher's performance in the classroom and teaching methodologies - 2013.

Proposition	Always (%)	Often (%)	A few times (%)	Never (%)	None (%)	More than one (%)
Does the teacher explain how mathematical knowledge has developed throughout history?	1,8	20,0	58,2	20,0	-	-
The teacher fully reproduces what is taught in the textbook.	3,6	50,9	41,8	3,6	-	-
The teacher teaches mathematical concepts according to the student's reality, presenting examples and problems related to their daily lives.	9,1	49,1	40,0	1,8	-	-
The teacher goes through and corrects lists of exercises in order to grasp the content.	23,6	52,8	21,8	1,8	-	-
Are the students shown various examples and illustrations so that they can better understand the content?	21,8	38,2	32,7	7,3	-	-
Can the teacher identify your difficulties and see if you really understand the subject?	10,9	43,7	38,2	5,5	1,8	-

The teacher encourages students to participate in class.	12,7	49,1	34,5	3,6	-	-
When a student asks a question that isn't related to the content being taught, the teacher doesn't go into the question and says that this subject will only be studied later or next year.	7,3	25,5	58,2	5,5	3,6	-
The math teacher is patient when explaining topics that seem difficult to understand.	25,5	41,8	23,6	7,3	1,8	-
The teaching materials provided by the teacher (handouts, books, copies, websites, etc.) help us to understand the subject.	12,7	49,1	30,9	7,3	-	-
Math lessons become more interesting when the teacher uses other resources such as videos, games, computers, etc., varying the style of the lesson.	45,5	21,8	18,2	12,8	1,8	-
It's nice when the teacher teaches "tricks" so that the student remembers how to use mathematical formulas.	40,0	38,2	9,1	-	12,7	-
It's important to check what the student already knows about a particular subject so that the teacher doesn't repeat too much of the same content.	32,7	34,5	25,5	5,5	1,8	-
Math class becomes more interesting when the teacher can relate it to other subjects.	18,2	41,8	38,2	1,8	-	-
Encouraging students to take part in championships, olympiads, lectures or fairs involving the teaching of mathematics helps to awaken their interest in this subject.	14,6	45,5	25,5	12,7	1,8	-
The student's interest in mathematics is stimulated when the teacher develops activities outside the classroom, such as laboratories,	29,1	32,7	27,3	10,9	-	-
sports courts, visits to other places, etc.						
Much of what I learn seems to have no meaning in my day-to-day life.	27,3	25,5	30,9	16,4	-	-
When I can actively participate in the lesson, asking questions and giving examples, I enjoy the lesson more.	41,8	36,4	20,0	1,8	-	-

Important points made by the students were that more than 75% of the teachers correct the lists of exercises, which is a very common way of learning, but the most important thing is the monitoring and attention that the teacher gives to this very important issue.

We also noticed the importance of encouraging students to take part in championships, as 60% indicated that this was a good influence on learning.

Chart 6. Perception of students in the 3rd year of secondary school regarding assessment practices - 2013.

Proposition	Always (%)	Often (%)	A few times (%)	Never (%)	None (%)	More than one (%)
The class receives guidance on how to study more efficiently.	9,09	29,09	50,92	7,27	3,63	
Does your teacher only prepare you to take his exam, or is he concerned with preparing you to transform your life, the environment in which you live and to face the problems of the world?	23,63	36,36	27,29	9,09	3,63	
The teacher's comments on your performance help you to improve the way you study and learn.	27,27	36,36	29,10	7,27		
The teacher has a habit of giving a more difficult test than the exercises done in class.	49,09	36,36	12,74	1,81		
The teacher adopts alternative assessment methods (individual and group work, reports, surveys, etc.).	25,47	36,36	36,36	1,81		

If we compare this item with the indications made, we can see some discrepancies between the data: in the first proposition, less than 39% of the students indicated that they were given the correct guidance on how to study. Another weak point, which we should emphasize, was the second topic, where almost 60% of the indications were that the teacher teaches only for the exam, a real paradigm to be broken by all of us teachers.

In item four, the students pointed out a flaw in the preparation of the tests, indicating that more than 80% of the students marked that the teacher had a habit of giving tests that were more difficult than the exercises, showing a clear lack of preparation in terms of the means of assessment.

With regard to alternative assessment options, around 61% said they had other assessment options, but perhaps, as we said earlier, they don't understand the reality of the students and the difficulties they face when it comes to assessment. This makes it clear that high school students need a little more attention than they currently get, because we can't consider them to be students prepared to face the world; the eighth grade generation is used to it and doesn't feel this lack of attention as much, if we compare the data.

5.3 Math teachers' perceptions

With the data obtained from the interviews with the teachers, we will begin a systematic analysis, where, out of respect for the teachers, as agreed in the interview, their names will not be disclosed, but we will call them by code names, A, B, C and so on. The interview script can be found in Appendix C, where we tried to get some baseline data, trying to find out where we are starting from, so that we can better guide our analysis.

The interview was carried out with the consensus of the teacher and the school, but for various reasons we didn't reach 100% of the teachers interviewed.

To explain a little more and make it clear, the interview was conducted by researcher Cleomar, who guided the interview questions for the interviewee.

We will analyze the six questions separately, but in the analysis of each question we will consider the answers of all the interviewees, within the analysis of each question and thus also be able to expose the ideas of the interviewees.

The first question is as follows. *What difficulties do you encounter in making teaching and learning effective?* In a question that dealt with two essential issues, teaching and learning, teacher A put the main barriers to this happening as lack of interest, demotivation, and a great lack of reflection on the part of the students on the part of the information in the world and also in the school. She mentions that '*everything around becomes more interesting than school subjects',* in other words, education competes with other sources that distract and take away attention, and she also mentions the large number of students in the classroom and the large amount of content.

In addition, teacher B mentions the *students' commitment to the school,* which is still not enough to achieve the results he wants. Teacher C, tells us that the *greatest difficulty is the lack of power to develop logical reasoning in sequence*, so that they can develop the mathematical concepts that he uses a lot, everything that has already been seen previously following the idea of sequence. The failure to develop this reasoning prevents students from uniting the concepts and mentions that '*students think they only need to know it for the test, and that's it'. He* also mentions the lack of study and the failure to prepare tasks outside the classroom as determining factors that prevent the healthy development of teaching and learning, he also talks about the bridges that teachers must create in order to achieve their teaching objectives, creating, as he himself mentions in the interview, '*bridges to link everyday mathematics with computers and new technologies, and applying classroom mathematics through these means... '*.

Working in a different way and thus gaining greater attention from the students, together with greater family participation, within education, he cites the fact that nowadays everything is ready-made, and cites television as the greatest disseminator of the idea of not reflecting or thinking about things, because there everything is ready to be consumed by our eyes and ears. These issues directly affect students' learning, which is a time when they have to reflect, think and come up with new ideas in order to build new knowledge. For teacher D, talking about the large number of students in the classroom, which also prevents the teacher from carrying out the work with quality, as she herself mentions, *because it is impossible for you to work with 40 students, 35 students in primary school, 40 in secondary school. If you're on your own with students who have a lot of difficulty, it's impossible for you to cover enough ground for everyone to be reached.*

In addition, most students have not developed the ability to interpret, which also ends up making things even more difficult, because if the exercises are different from those proposed in the examples, the students are unable to develop their own methods of solving them. For teacher E, all of the above is correct and she also adds that *in addition to all this difficulty, there is a lack of encouragement from her family and from the students themselves, who don't give a damn about it. So we, as teachers, try, captivate them and go and look for them, but they simply won't come.* The problem has already contaminated the whole system; it's hard to get good students from schools that don't keep up with reality and current affairs. As she herself says, *what matters to them [the government] is quantity, statistics, and to us, unfortunately for us, it's quality. So we lose out because of this.* They are no longer concerned with the quality of teaching, but with the number of students who pass, which fattens the government's paltry statistics. For teacher F, the lack of respect between the students and the teacher is also a problem that affects a lot, creating difficulties for the student's other classmates. She mentions that *for the teacher to be able to explain, to be able to perform well in their area, regardless of the subject, there has to be respect on both sides.*

We can see that there is also a lack of cooperation from the students and also from the family, because respect is something that we acquire through living together, seeing and hearing it used in our daily lives, having respect for teachers and colleagues who want to learn, but who, due to the lack of interest of some, end up getting in the way of the whole class.

In the second question, we asked the following question. *What are the characteristics of a "good" math teacher?* Teacher A said that *a good math teacher should be patient and explain things in a different way, trying to reach all students. In addition, they should use everyday situations to explore mathematical concepts.* We can see that these are several attitudes that a teacher must have in order to be a good educator, because by combining patience with comprehensive and well-developed explanations, they can win over all the students and focus their attention on what really matters. For teacher B, a good teacher must know how to listen to the students and also work on this, explaining the reason for mathematics, the development it needs to become more understandable.

For teacher C, the good teacher is the one who contextualizes his lessons, works on content that is directly linked to everyday life, and also works with reality, thus teaching ideas that are more present in everyday life, a classic example he cites is '*I particularly teach fractions together with economics, I work with the value of the salaries of these students' parents,* He also says that the good teacher studies the class and identifies the problems that exist in it in order to be able to come up with better solutions to overcome these deficiencies, saying that each class is different and it is necessary to study all the classes in order to really be able to build knowledge.

The teacher must know how to work with the content, interconnecting various areas of knowledge, how to work in a context that is outside the classroom, bring in the technologies that are present and elaborate and analyze the data through computer software that works with mathematics and its properties, creating in the student a new perspective on the world and even a sharper understanding of the problems experienced in today's world. In short, the good teacher, as he himself says, '*must be responsible for creating and bringing a dynamic, fun and profitable lesson.* Breaking the routine and paradigms that mathematics is a boring lesson, transforming the subject itself into an attraction for the students.

For teacher E, in order to have the minimum conditions to be able to indicate what a good teacher is, we need conditions such as basically *having more materials available that we don't have, availability of material, practical things that we could be using, that we don't have, to* be able to work in a differentiated way, and attend to students at a much higher level of satisfaction, to have these conditions so that playful activities are more frequent for students.

For teacher D, as she herself says, it *would be good not to work only with the traditional: board, books, activities, play a game, take the students to the computer room, use the data show; we even have this in the school, but it's difficult to adapt it to the classroom, because there isn't even enough computer for a whole class, there's* a lack of structure for us to work harder too, and thus reap more rewards, we realize that the teachers do want to be the difference, but the basics are lacking, and so we are stuck with the books and the classroom. For teacher F, the characteristics are diversity in working with the content, *and also inserting other ways, other activities, working in various ways, games, magic, videos and so on*, breaking the routine, and leaving the student always motivated and curious to know what they will have in the next class.

In the third question, we are already looking to engage with innovation, to find out how this very important topic is doing. What innovations has the school experienced in recent years? According to teacher A, schools are slowly starting to introduce new and up-to-date things. We can conclude this from what she says in the interview: "*In recent years, the school has had science and geography laboratories, as well as an excellent computer room. There has also been an increase in the number of lectures, theaters and courses, in other words, a conversation with reality and the social environment.* In a nutshell, he stressed the importance of transferring knowledge in a more differentiated way, making lessons more attractive to students.

Another essential point is the technological quality that schools have been receiving, which also influences the self-esteem of both the students and the teachers themselves, with the possibility of being able to work in a different way and diversify lessons with work that shows mathematics through the eyes of technology. For teacher B, *the school should also have a physical space where teachers and students can prepare for math tests, or together with chemistry and physics have a complete laboratory with space, material and teachers prepared to work in this place*, because innovation means diversifying and getting out of the routine of the classroom, books, tests and assignments, we can work with mathematics from a different angle, being able to see it transforming with the help of technologies.

For teacher C, there is a lack of policies to encourage innovation in schools, both in mathematics and in other areas. The teacher says that research would be the greatest innovation to be implemented in schools, but for this to happen we need someone to support this idea and to pay for these studies, because producing knowledge is much more important than just reproducing it, our students don't know what scientific or educational research really means.

If we teachers had the time to be able to work with these resources, we wouldn't need to "import" knowledge so urgently, we'd be producing it right here, but our governments prefer to buy only material things, instead of investing in projects and collecting results after the end of the research. There is a lack of incentive for students and teachers to do research. And so we stand still in time, while other countries get ahead of us.

For teacher F, the changes and innovations we are experiencing are still slow and are mainly linked to *technology, the issue of computer labs. The school I work at has a computer lab, there are even two computer labs, and a science lab. These are things that they didn't have until a few years ago.* As the teacher told us, this is not enough for anyone who thinks of being a developed country. Innovation goes far beyond technology, it can also be introduced by the teacher's pedagogical and didactic actions. Even within assessments and the development of knowledge itself, such as research, which is a very essential tool if used wisely, and to break the paradigms of schools and advance towards the future in leaps and bounds.

For teacher E, mainly the installation and maintenance of the computer room, data show, multimedia rooms, with DVDs. Which are almost impossible to use, because there are so many teachers wanting to use them, and so little equipment, as well as the tablets not being able to connect to the data show, as the teacher says *I have one, I don't use it, it's for secondary school teachers, I even got mine. But they didn't give us instructions on how to use it, it doesn't come with a text editor, if you want to use it in a data show, it doesn't have the capacity for that, it doesn't have an input to connect this media, so why?*

The question remains as to what the intention really is, to introduce an innovation without first preparing the teachers and even the students to receive this technological innovation. The teacher is indignant when she talks about how she is going to work with 40 students and still try to master the functions of the equipment at the same time. Teacher D, on the other hand, sees the inequalities there, as only teachers who are permanent at secondary school have received this technology, and *have implemented tablets in the school. Well, there are the tablets, but who's to say that everyone knows how to use them? how to use the tablet in the classroom?*

It would be nice if everyone had their own tablet, teachers and students, so that there could be this interaction in the classroom with this technology, which could bear a lot of fruit, as it would bring a little more joy to the students, as well as greater motivation.

As we saw earlier, we now want to know '*What innovations can you, as a math teacher, incorporate into your lessons?* For teacher A, it is possible to work in my classes '*videos, computer room, use of software and games*, if we stop to think it is very little, for those who seek to prepare students for a lifetime, but these are the realities faced by teachers, each day facing a new challenge, and always seeking to overcome and learn lessons, and show students that we do not have everything we want at our disposal, and yet we seek to give our best.

For teacher B, the lack of time often means that the school doesn't keep up with new things like the math Olympiad. Having the time to prepare for something different would be attractive for the students, because the teacher often ends up holding back, always having to schedule time in computer labs, and also leaving aside the idea of mixing and preparing an innovative and differentiated lesson.

For teacher C, the use of technology is very important, because it shows the student that the teacher prepares and organizes himself to be bringing the best to the students, the teacher mentions some equipment that he uses which are *the digital whiteboard, software, computer room for research,* I prepare my classes at home with the help of the software that we have at our disposal and I take my data show, my notebook, and I work in the classroom with the help of technology that shows the students, how, when and where we use each content of mathematics, He also comments on the lack of preparation for handling the technologies that already exist, because not everyone knows how to prepare a lesson and apply it on the digital whiteboard, there is also a lack of guidance on how to proceed with this development work, the steps to create an attractive and differentiated lesson that really catches the students' attention, because nowadays, as the teacher mentions, *'everything is more attractive than the classroom'.*

For the teacher, innovation is a tool that is available to complement and give the 'final touch', after having all the foundations in the classroom, but following an order, in the construction of knowledge, as he tells us. Start in the classroom with all the knowledge, apply exercises, show theory, and practice, so that the student can adapt to the concepts, and possibly then introduce the tools of innovation, because the teacher has a duty to alleviate the problems in the classroom, which are not few, so that there can be real learning.

For teacher E, when asked, she responded thoughtfully, saying *Very little. Very little because there's no way we can do much that's different.* There's a lack of structure and ideas, of being able to prepare something outside the norm. We see how important it is to have ideal conditions, time and structure to present the content in a different way.

For teacher F, everything has its moment, *depending on how I'm working, currently it's with the seventh grades, at certain times it's not possible, but always when there are subjects where it's possible to use the computer lab, we end up taking the students. It has internet access, it's very easy to use, it*'s a question of adapting and planning ahead so that things work. Organization is also essential if we are to be able to take innovation from planning to action.

The fifth question asked: *What is essential in a classroom for the teacher to be able to teach and the student to be able to learn?* For teacher A, an *adequate space that allows total concentration is essential. It is also essential for the teacher to be able to teach what they know, in other words, to have teaching techniques.* In other words, it's about mastering the content first and foremost, a suitable location, and students who are focused and really concentrating on the lessons. It's said that there are many things more attractive than the classroom these days, but as the teacher tells us, it's up to us teachers to make our lessons truly attractive for the students, to transmit that satisfaction, that joy of being in the classroom, is also part of the indispensable context for the student to be able to learn, motivation is also learning.

For teacher B, - *willingness on both sides, and a well-prepared, optimistic teacher and, on the other hand, a student who is ready to learn, and willing to listen and interact and not hold back.'* Also complementing what the teacher told us earlier, motivation and willpower, and being prepared for challenges.

For teacher C, he talks about balance, because everything has to be dosed in the right way, to get to the essential point, which is the students' learning. On this issue, everyone has similar ideas about what is essential in a classroom for learning to take place.

For teacher F, what has become indispensable these days is respect, as has already been mentioned, respect between both teachers and students, so that everything runs as smoothly as possible, without interventions that could take away the students' attention and concentration.

For teacher D, having *these innovations out there, suggestions on how to use them within Mathematics,* can be a differentiator, so that teaching and learning are completed at the same pace and everyone can learn from the teacher's words, ideas and attitudes. Another essential point is the lack of resources, as each student was initially entitled to five pages per bimester in copies, but for various reasons, this number has been further reduced, meaning that the teacher still spends a lot of time going over the board, while this time could be used to work on new things, but we are left without action in the face of these concerns, for the teacher *it would be basic teaching material that the school cannot provide because it does not have the funds for it.*

We're still a long way from perfect schools, our infrastructure is practically nowhere to be found, but it's everyone's duty to demand free, high-quality materials as a minimum.

Teacher E talks about the responsibility of researching teachers before sending them any material: "*We have a Lego game here at school, boxed up, because we don't even know what it's for, what it's used for. There are lots of games that can be used in mathematics, but we don't know how to use them. So, first of all, before spreading a bunch of stuff in the media, before wanting to buy things that someone has produced and wants to sell, they should consult the teachers, who are there in the classroom on a daily basis and check with them: we have the idea of providing material, so what kind of material would we need? but for this material to work it needs a course, it needs improvement. How can we, all of a sudden, create something together, develop it together*, let's take action as far as possible so that we can improve performance inside and outside the classroom, through materials, research, a group of factors that influence how well things work.

And to complete the interviews, we asked, "*Do you have any suggestions for our research?"* Teacher A said in the interview, "*Thinking of suggestions to make teaching more efficient"* - words that say a lot, especially the realization of knowledge, as a way of turning ideas into actions, making things happen, because we are full of plans and projects, what we lack are attitudes and actions that really aim to solve problems.

For teacher C, he suggests that we seek a closer relationship between the school and the university, to bring more results in relation to research, because research always brings us something good within the school itself, bringing results and new things to be analyzed to change some paradigms in relation to scientific and even educational research, The researcher must be committed to bringing the results and showing them to us in the form of a report or lecture, so that we are aware of what is really happening within the school and with our students.

CONCLUSIONS

Based on the results of this survey, it is possible to see that a series of factors that are linked to other factors of paramount importance in the construction of knowledge, having a good teacher is already halfway to knowledge, making life much easier for students, and now we can outline the characteristics of a good teacher. When analyzing the students' perceptions of this issue, it can be seen that some of these factors are more recurrent, such as the teacher's behavior and attitudes, which are evident in the students' expectations. This study clearly reveals the link between a good math lesson and a good teacher, i.e. when the relationship between student and teacher is good, this has a direct impact on performance in the classroom.

Starting with an in-depth analysis, we can also say that schools have been under pressure from society, parents, principals and the system itself, which wants to work mainly in high school with a very strong focus on entrance exams for graduates from large universities, leaving aside the ideas set out in the PCNs, and in the educational guidelines, as Krasilchik (1995), tells us that this has generated major changes in teaching structures, as the school has sought to meet at least the criteria to reach a common denominator.

The entrance exams also represent a considerable external pressure on secondary education. The programs presented by the big universities or the institutions that organize the entrance course have a normative effect on teaching. In addition to the courses and colleges that set out to prepare students for the entrance exams, families and students themselves put pressure on schools to meet the university entrance exams. (KRASILCHIK, p. 192, 1995).

The transformation of teaching must seek to satisfy several areas, it must cover educational satisfaction, create a subject that fits into our society, that really participates in everyday life, that is, that our teaching system of innovation, differentiated practices, the good teacher and teaching methods mixed with technology, the content being geared to the social reality of the students. Concrete and positive results will please students, teachers, businesspeople and professional society alike.

When the teaching of science is based on teaching methods and seeks to make teaching practical, attractive and to give students the opportunity to get to know scientific work, and the educational community in a convinced way, that makes the effort worthwhile, thus leaving a much more consistent legacy.

We would like to highlight here the chapter that emphasizes the characteristics of a good teacher, from which we can see from various analyses that students understand the content through different actions. We have tried to emphasize in a general way, from renowned authors, which of these characteristics have already been raised, and which still prevail today, of which we have chosen the ten commandments of the good teacher, reported by Polya (1959), which summarizes very well the essential characteristics of a good teacher.

In order for teaching to take place, we have given a brief overview of teaching methods, because they are often not considered important, but it is the methods that are responsible for making knowledge effective, and it is through them that learning takes place, because methods are the organizational structures that aim not only at learning but also at the physical well-being of the students. All the elements are essential, such as teaching didactics, teacher qualification and the pedagogical practices applied continuously, which are of the utmost importance for effective teaching and learning.

When it comes to pedagogical innovation, we come across a wide range of tools, objects and actions, and it's often not enough just to have the technology if we don't have the ideal training to work with it. We have learned that innovation comes with planning, organization, investment and training. To innovate is much more than transforming, but maintaining and reinventing day-to-day life in the classroom, it is knowing how to work in a different way, and building knowledge on top of tools such as digital whiteboards, data shows, computer software, among other elements, which will certainly contribute significantly to student learning.

The use of technology as a simple and practical form of innovation, combining virtual whiteboards with lessons planned using computer systems, makes them more productive and leads to a greater understanding of certain content that is abstract in theory.

In the chapter on methods and procedures, we explain how the research was carried out and what our objectives were, which were to identify the characteristics of a good teacher, starting with the application of the questionnaire to the students. We also analyzed the interviews with the math teachers, from which we were able to draw a lot of important information that is explained in Chapter Five. In this chapter we can see and conclude that the schools surveyed still do not have the full conditions in which to work, because there is a lack of investment in structure, but the most important thing is to see that the teachers and principals work within this environment and still manage to achieve good results.

In the chapter on the presentation and analysis of data, perhaps the most important of these stages, we can reach extremely important conclusions through our analysis, even breaking paradigms. In my opinion, I had always believed that the vast majority of students didn't like mathematics, but our research proves that in the context of our population, more than 53% of students like mathematics, which was directly reflected in the indication of who was a good teacher, where the most indicated were mathematics teachers. This shows that we are already deconstructing the idea that mathematics is difficult, and we came to the conclusion that students in the eighth grade indicate good teachers because of their personal characteristics.

The students in the third grades of secondary school, on the other hand, indicated that they were more impressed by the teacher's characteristics in the classroom, such as explaining well, mastering the content, knowing how to assess, and explaining in a way that everyone can understand. It will be from understanding this research that we will be able to improve our performance in the classroom as future teachers.

When this research was planned, we set ourselves some goals, such as where we wanted to go and what we wanted to achieve. In the end, we are generally satisfied with the results achieved and presented here. The part of collecting the data was perhaps the most tiring and exhausting, facing many difficulties such as the teachers' lack of time, interest and commitment. They would make an appointment, get there and forget, have a test or an assignment, and not be able to give the interview, but in spite of everything we covered as many teachers as possible, six out of a total of eleven, which also resulted in a lot of data that was analyzed, pointing to a lack of infrastructure, government commitment, conditions and even the will of the students.

To complete this cycle, I would like, if possible, to extend this research to get to know more of the university professors, because we know the characteristics of those who train the individuals who enter university, and now I would like to get to know the professors who train the individuals who lead our society, thus becoming a new subject for future research. I would like to express my opinion, which I wholeheartedly agree with, on the results obtained from the students' notes and the teachers' ideas and recommendations.

I hope I have aroused interest among all those involved in this research, students, teachers, principals and advisors. I hope I have ignited a flame of hope within teachers, awakening the desire even in their classmates to become a good teacher, given the characteristics I have listed in my research.

REFERENCES

AURICH, Grace Da Ré. **Games of truth in the constitution of the good math teacher**. 2011. 117 f. Dissertation (Master's in Education) - Faculty of Education, Postgraduate Program in Education, Federal University of Rio Grande do Sul, Porto Alegre, 2011. Available at: <http://hdl.handle.net/10183/36395>. Accessed on: July 5, 2012.

Brazilian advances in basic education. Available at: http://www.youtube.com /watch?v=pTx-vmHjsHs&feature=youtu.be. Accessed on: July 6, 2012.

CUNHA, Maria Isabel da. The relationship between teaching and research. In: VEIGA, Ilma Passos de Alencastro (Org.). **Didática: o ensino e suas relações**. 6.ed. Campinas, SP: Papirus, 1996, p. 115-126.(magistério:formação e trabalho pedagógico).

CUNHA, Maria Isabel da. The teacher-student relationship. In: VEIGA, Ilma Passos de Alencastro (Org.). **Repensando a didâtica**. 5.ed. Campinas, SP: Papirus, 1991, p. 145-158.

CUNHA, Maria Isabel da. **The good teacher and his practice.** 11.ed. Campinas, SP: Papirus, 1989.(Teaching: Training and pedagogical work).

CUNHA, Maria Isabel. **Pedagogical innovations**: the challenge of reconfiguring knowledge in university teaching. Sao Paulo: USP, 2008 (Cadernos Pedagogia Universitària).

DEMO, Pedro. **Education, Qualitative Assessment and Innovation**. Quality of education. Brasilia: Instituto Nacional de Estudos e Pesquisas Educacionais Anisio Teixeira, 2012.(Série Documental. Textos para Discussao).

GARCIA, Walter Esteves, (coord.). **Inovaçâo educacional no Brasil:** problemas e perspectivas. Campinas: Editora autores associados, 3ª Ed. 1995.

GARCIA, Walter Esteves. Legislation and educational innovation since 1930. at. In.

GARCIA, Walter Esteves, (coord.). **Inovaçâo educacional no Brasil**. campinas: Editora Autores Associados, 1995. p. 223-254.

GRUPO RBS. **Education needs answers**. Available at: <http://www.clicrbs. com.br/ especial/br/precisamosderespostas/ pagina,1428,0,0,0,Sobre-o-projeto.html>. Accessed on: 05 Dec. 2012.

HERNANDEZ, Fernando et al. **Aprendendo com as inovaçôes nas escolas;** trad. Ernani Rosa. Porto Alegre: Artes Médicas Sul, Artmed, 2000.

KRASILCHIK, Myrian. Innovation in science teaching. at. In. GARCIA, Walter Esteves, (coord.). **Inovaçao educacional no Brasil**. campinas: Editora Autores Associados, 1995. p. 177-194.

MACEDO, Lino de. Situaçao problema: forma e recurso de avaliaçao, desenvolvimento de competências e aprendizagem escolar. at. In: PERRENOUD, Philippe (Org.). **Teacher training in the 21st century**. Porto Alegre: Artmed, 2002. p. 113-136.

MACHADO, Nilson José. On the Idea of Competence. at. In: PERRENOUD, Philippe (Org.). **A formaçao dos professores no século XXI**. Porto Alegre: Artmed, 2002. p. 137156.

MOVIMENTO Todos pela Educaçao. **A good teacher, a good start** (Jingle).

Available at:<http://www.todospelaeducacao.org.br/comunicacao-e-midia/pecas-de-

cominicacao/videos/509/um-bom-professor-um-bom-comeco>. Accessed on: June 23, 2011.

MOYSÉS, Lucia. **The challenge of knowing how to teach.** 5. ed. Campinas, SP: Papirus, 1994.

OLIVEIRA, Rosiele Juvino de. **The good math teacher according to the perception of high school students.** 2007. Final course work (Mathematics), Catholic University of Brasilia, Brasilia, 2007. Available at: <http://www.ucb.br/sites/100/103/TCC/ 12007/RosieleJuvinodeOliveira.pdf>. Accessed on: June 6, 2012.

PERRENOUD, Philippe. **Differentiated pedagogy**: from intentions to action. Porto Alegre: Artmed, 2000.

PÓLYA, George. Ten Commandments for Teachers. **Revista do Professor de Matemàtica**, Sao Paulo, n.10, p. 2-10, 1987.

RAYS, Oswaldo Alonso. The Theory-Practice Relationship in Critical School Didactics. at. In: VEIGA, Ilma Passos de Alencastro (Org.). **Didâtica: O Ensino e Suas Relaçôes**. 6.ed. Campinas, SP: Papirus, 1996, p. 33-52.

RAYS, Oswaldo Alonso. Teaching methodology: culture of the contextualized path. at. In: VEIGA, Ilma Passos de Alencastro (Org.). **Repensando a didâtica**. 5.ed. Campinas, SP: Papirus, 1991, p. 93-104.

RIZZOTTI, Maria Angela. A good math teacher should...: students' representations of teacher mediation. **Ensenanza de las Ciencias**, n. extra, 2005.

THURLER, Monica Gather. From teacher evaluation to school evaluation. at. In: PERRENOUD, Philippe (Org.). **Teacher training in the 21st century**. Porto Alegre: Artmed, 2002. p. 61-112.

WACHOWICZ, Lilian Anna. **Mediating Pedagogy**. Petrópolis, RJ: Vozes, 2009.

APPENDICES

APPENDIX A - Questionnaire applied to students

COMMUNITY UNIVERSITY OF THE CHAPECÓ REGION

COURSE: Mathematics

CURRICULAR COMPONENT: Research ii

RESEARCHER: **Cleomar Alexandre Hirt**

GUIDING PROFESSOR: Clàudia Maria Grando

QUESTIONNAIRE

You are being invited to take part in a survey. *The aim of this questionnaire is to survey the characteristics of mathematics teachers in order to identify which of them are necessary for a "good teacher".* We ask you to be very honest and inform us that we will not use your name in the survey, only the information collected. Thank you in advance for your cooperation.

Identification data

Name (optional):

Sex: ( ) Male ( ) Female:

Name of school:

Series: .... Class:

Open questions

1. Did you have a special teacher who made a difference in your life?
2. What subject did this teacher teach?
3. Why was he special to you?
4. What differentiates a good teacher from other teachers?
5. Who was the best math teacher you ever had? In which grade(s) did they teach you? (You can name more than one if you like)
6. What characteristics should a "good teacher" of mathematics have?
7. What should a "good lesson" in mathematics be like?
8. Do you like mathematics? Why do you like it?

Questions to mark: Read each proposition and mark with an "x" the alternative that best suits you, in general, or with regard to your current math teacher.

Proposition	Always	Often	A few times	Never
Teachers who have a love of mathematics are able to make students enjoy learning it.				
A teacher who doesn't relate well to the students makes them enjoy the class less.				
An attentive and friendly teacher encourages learning.				
Teachers who are enthusiastic about their subject motivate me to study more.				
When you can't learn the content, do you just memorize it?				
Does the teacher explain how mathematical knowledge has developed throughout history?				
The teacher fully reproduces what is taught in the textbook.				
The teacher teaches mathematical concepts according to the student's reality, presenting examples and problems related to their day-to-day lives.				
The teacher goes through and corrects lists of exercises in order to grasp the content.				
Various examples and illustrations are presented to the students so that they can better understand the content?				
Can the teacher identify your difficulties and see if you really understand the subject?				
The teacher encourages students to participate in class.				
When a student asks a question that isn't related to the content being taught, the teacher doesn't go into the question and says that this subject will only be studied later or next year.				
The math teacher is patient when explaining topics that seem difficult to understand.				
The students' opinions are valued by the teacher.				
I know I can count on my teacher even outside the classroom.				
The class receives guidance on how to study more efficiently.				
Does your teacher only prepare you to take his exam, or is he concerned with preparing you to transform your life, the environment in which you live and to face the problems of the world?				
The teacher's comments on your performance help you to improve the way you study and learn.				
The teacher has a habit of giving a more difficult test than the exercises done in class.				

The teacher uses alternative assessment methods (individual and group work, reports, surveys, etc.).				
The teaching materials provided by the teacher (handouts, books, copies, websites, etc.) help us to understand the subject.				
Math lessons become more interesting when the teacher uses other resources such as videos, games, computers, etc., varying the style of the lesson.				
It's nice when the teacher teaches "tricks" so that the student remembers how to use the mathematical formulas.				
It's important to check what the student already knows about a particular subject so that the teacher doesn't repeat too much of the same content.				
Math class becomes more interesting when the teacher can relate it to other subjects.				
Encouraging students to take part in championships, olympiads, lectures or fairs involving the teaching of mathematics helps to awaken their interest in this subject.				
The student's interest in mathematics is stimulated when the teacher develops activities outside the classroom, such as laboratories, sports grounds, visits to other places, etc.				
Much of what I learn seems to have no meaning in my day-to-day life.				
When I can actively participate in the lesson, asking questions and giving examples, I enjoy the lesson more.				

APPENDIX B - Interview script

INTERVIEW SCRIPT

1. **Identification Data** (the name of the teacher will not be disclosed in the Research Report, the data will be worked on in general)

Sex: Age: .... Length of time teaching: ....

Academic Background: - Degree in ....

( ) Complete ( ) Incomplete

- Postgraduate ( ) Specialization in...

( ) Master in

( ) Doctorate in

School(s) where you work: ....

Grades/years in which you teach: ....

Level: ( ) Elementary ( ) Middle ( ) Superior

Weekly workload:..

Do you have another job? ....

2. Interview (ask permission to record the interview)

a) What difficulties do you encounter in making teaching and learning work?

b) What are the characteristics of a good math teacher?

c) What innovations has the school experienced in recent years?

d) What innovations can you, as a math teacher, incorporate into your lessons?

e) What is essential in a classroom for the teacher to be able to teach and the student to be able to learn?

f) Do you have any suggestions for our research?

APPENDIX C - Consent form for data collection

COMMUNITY UNIVERSITY OF THE CHAPECÓ REGION

AREA OF EXACT AND ENVIRONMENTAL SCIENCES - ACEA

COURSE: Mathematics

RESEARCH PROJECT: "Innovative practices of the 'good teacher' of mathematics"

RESEARCHER: **Cleomar Alexandre Hirt**

GUIDING PROFESSOR: Clàudia Maria Grando

INFORMED CONSENT FORM

This school is being invited to participate as a volunteer in a study. Once you have been informed of the following information, if you agree to take part in the study, please fill in the final details and sign the document.

The research is entitled *Innovative practices of the "good teacher" of mathematics* and is being developed by the academic Cleomar Alexandre Hirt of the 7th period of the Mathematics Course at Unochapecó and supervised by Professor Ms. Clàudia Maria Grando. We are sending you a copy of the research project which contains all the details.

Data collection will consist of the use, without any harm or embarrassment to the respondent, of a questionnaire with open and closed questions applied to students in the 9th grade (or 8ª) of elementary school and also in the final year of secondary school in the schools of Pinhalzinho - SC, the city of residence of the student researcher. We will also use interviews with smaller groups of students to expand on the positions presented. The interview will also be used with teachers who teach mathematics, to identify their point of view on their teaching practice, the difficulties encountered in teaching and learning and the characteristics of a good mathematics teacher. A copy of the questionnaire is attached for your analysis. The information obtained through data collection will be used to answer the research problem and questions and to compose the research report.

We count on your important contribution to the professional training of student researchers and to promoting discussion and improving the teaching of mathematics in our schools.

We would like to thank you and would be happy to answer any questions you may have by calling (49) 3321 8024 and (49) 8854 2734, or by e-mailing claudia@unochapeco.edu.br and cleohirt@unochapeco.edu.br.

CONSENT TO PARTICIPATE

I, , ID ______________________________________

I, the undersigned school manager , agree to this educational institution taking part in the above-mentioned research. I have been duly informed and clarified about the research, the procedures involved, as well as the benefits arising from my participation. I am also assured that I can withdraw my consent at any time without any penalty and that the identity of those involved in the research will be preserved.

Place: ______________________________________ Date: ___/___/ _________

Name: ____

Signature: ____

MIX
Papier aus verantwortungsvollen Quellen
Paper from responsible sources
FSC® C105338

Printed by Books on Demand GmbH, Norderstedt / Germany